REA's Books Are

They have rescued lots of ~~~~~~~

"Your books are great! They are very helpful, and have upped my grade in every class. Thank you for such a great product."
Student, Seattle, WA

"Your book has really helped me sharpen my skills and improve my weak areas. Definitely will buy more."
Student, Buffalo, NY

"Compared to the other books that my fellow students had, your book was the most useful in helping me get a great score."
Student, North Hollywood, CA

"I really appreciate the help from your excellent book. Please keep up your great work."
Student, Albuquerque, NM

"Your book was such a better value and was so much more complete than anything your competition has produced (and I have them all)!"
Teacher, Virginia Beach, VA

(more on next page)

" Your books have saved my GPA, and quite possibly my sanity. My course grade is now an 'A', and I couldn't be happier. "

Student, Winchester, IN

" These books are the best review books on the market. They are fantastic! "

Student, New Orleans, LA

" Your book was responsible for my success on the exam. . . I will look for REA the next time I need help. "

Student, Chesterfield, MO

" I think it is the greatest study guide I have ever used! "

Student, Anchorage, AK

" I encourage others to buy REA because of their superiority. Please continue to produce the best quality books on the market. "

Student, San Jose, CA

" Just a short note to say thanks for the great support your book gave me in helping me pass the test . . . I'm on my way to a B.S. degree because of you ! "

Student, Orlando, FL

STATISTICS

By the Staff of
Research & Education Association
Dr. M. Fogiel, Director

Research & Education Association
61 Ethel Road West
Piscataway, New Jersey 08854

SUPER REVIEW™
OF STATISTICS

Printed in the United States of America

Library of Congress Catalog Card Number 00-132737

International Standard Book Number 0-87891-197-9

SUPER REVIEW is a trademark of
Research & Education Association, Piscataway, New Jersey 08854

WHAT THIS Super Review WILL DO FOR YOU

This **Super Review** provides all that you need to know to do your homework effectively and succeed on exams and quizzes.

The book focuses on the core aspects of the subject, and helps you to grasp the important elements quickly and easily.

Outstanding **Super Review** features:

- Topics are covered in logical sequence

- Topics are reviewed in a concise and comprehensive manner

- The material is presented in student-friendly language that makes it easy to follow and understand

- Individual topics can be easily located

- Provides excellent preparation for midterms, finals and in-between quizzes

- In every chapter, reviews of individual topics are accompanied by Questions **Q** and Answers **A** that show how to work out specific problems

- At the end of most chapters, quizzes with answers are included to enable you to practice and test yourself to pinpoint your strengths and weaknesses

- Written by professionals and test experts who function as your very own tutors

Dr. Max Fogiel
Program Director

CONTENTS

vii

CHAPTER 1

Introduction

1.1 What is Statistics?

Statistics is the science of assembling, organizing, and analyzing data, as well as drawing conclusions about what the data means. Often, these conclusions come in the form of predictions. Hence, the importance of statistics.

The first task of a statistician conducting a study is to define the population and to choose a sample. In most cases, it is impractical or impossible to examine the entire group, which is called the **population** or **universe**. This being the case, one must instead examine a part of the population called a **sample**. The chosen sample has to reflect as closely as possible the characteristics of the population. A population can be finite or infinite.

For example, statistical methods were used in testing the Salk vaccine, which protects children against polio. The sample consisted of 400,000 children. Half of the children, chosen at random, received the Salk vaccine; the other half received an inactive solution. After the study was completed, it turned out that 50 cases of polio appeared in the group that received the vaccine, and 150 cases appeared in the group that did not receive the vaccine. Based on that study of a sample of 400,000 children, it was decided that the Salk vaccine was effective for the entire population.

EXAMPLES:

The population of all television sets in the USA is finite.

The population consisting of all possible outcomes in successive tosses of a coin is infinite.

Descriptive or deductive statistics collects data concerning a given group and analyzes it without drawing conclusions.

From an analysis of a sample, inductive statistics or statistical inference draws conclusions about the population.

Problem Solving Example:

Suppose that a sociologist desires to study the religious habits of 20-year-old males in the United States. He draws a sample from the 20-year-old males of a large city to make his study. Describe the sample and target populations. What problems arise in drawing conclusions from this data?

 The sample population consists of the 20-year-old males in the city that the sociologist samples.

The target population consists of the 20-year-old males in the United States.

In drawing conclusions about the target population, the researcher must be careful. His data may not reflect religious habits of 20-year-old males in the United States but rather the religious habits of 20-year-old males in the city that was surveyed. The reliability of the extrapolation cannot be measured in probabilistic terms.

1.2 Variables

Variables will be denoted by symbols, such as x, y, z, t, a, M, etc. A variable can assume any of its prescribed values. A set of all the possible values of a variable is called its **domain**.

EXAMPLE:

For a toss of a die, the domain is {1, 2, 3, 4, 5, 6}.

The variable can be discrete or continuous.

EXAMPLE:

The income of an individual is a discrete variable. It can be $1,000 or $1,000.01, but it cannot be between these two numbers.

EXAMPLE:

The height of a person can be 70 inches or 70.1 inches. It can also assume any value between these two numbers. Hence, height is a continuous variable.

Similarly, data can be discrete or continuous. Usually, countings and enumerations yield discrete data, while measurements yield continuous data.

Problem Solving Examples:

 Discuss and distinguish between discrete and continuous values.

The kinds of numbers that can take on any fractional or integer value between specified limits are categorized as continuous, whereas values that are usually restricted to whole-number values are called discrete. Thus, if we identify the number of people who use each of several brands of toothpaste, the data generated must be discrete. If we determine the heights and weights of a group of college men, the data generated is continuous.

However, in certain situations, fractional values are also integers. For example, stock prices are generally quoted to the one-eighth of a dollar. Since other fractional values between, say, 24.5 and 24.37 cannot occur, these values can be considered discrete. However, the discrete values that we consider are usually integers.

Q (a) Suppose a manufacturer conducts a study to determine the average retail price being charged for his product in a particular market area. Is such a variable discrete or continuous?

(b) In conjunction with the previous study, the manufacturer also wants to determine the number of units sold in the area during the week in which an advertising campaign was conducted. Is this variable discrete or continuous?

A (a) Since an average may take any fractional value, the average retail price is continuous.

(b) We have a count, therefore, this variable must be discrete. The number of units sold is a discrete variable.

1.3   Functions

Y is a function of X if, for each value of X, there corresponds one and only one value of Y. We write

$$Y = f(X)$$

to indicate that Y is a function of X. The variable X is called the independent variable, and the variable Y is called the dependent variable.

EXAMPLE:

The distance, s, travelled by a car moving with a constant speed is a function of time, t.

$$s = f(t)$$

If $s = 16t^2$, we could write $f(t) = 16t^2$. Thus, for example, if $t = 2$, $s = f(2) = (16)(2^2) = 64$. There are also functions of two or more independent variables.

$$z = z(x, y)$$

$$y = f(x_1, x_2, \ldots, x_n)$$

The functional dependence can be depicted in the form of a table or an equation.

EXAMPLE:

Mr. Brown can present income from real estate in the form of a table.

Year	Income
1	18,000
2	17,550
3	17,900
4	18,200
5	18,600
6	19,400
7	23,500
8	28,000

EXAMPLE:

$$y = 3x + 2$$

The value of a variable y is determined by the above equation.

x	-3	-2	-1	0	1	2	3
y	-7	-4	-1	2	5	8	11

We will be using rectangular coordinates.

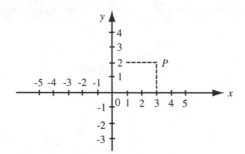

The coordinates of point P are $x = 3$ and $y = 2$. We write $(3, 2)$ to represent this.

Graphs illustrate the dependence between variables. This book discusses linear graphs, bar graphs, and pie graphs.

Problem Solving Examples:

 If $y = 4x - 3$, write the values of y corresponding to $x = 1, 2, 3, 4, 5$ in table form.

 $y = (4 \times 1) - 3 = 4 - 3 = 1$

$y = (4 \times 2) - 3 = 8 - 3 = 5$

$y = (4 \times 3) - 3 = 12 - 3 = 9$

$y = (4 \times 4) - 3 = 16 - 3 = 13$

$y = (4 \times 5) - 3 = 20 - 3 = 17$

x	1	2	3	4	5
y	1	5	9	13	17

 Using the letter g, represent symbolically the statement that D is a function of k.

 $D = g(k)$. The variable on the left side of the equation, D, is a function of (that is, it varies with) the variable in parentheses on the right side. The letter g indicates "a function of," so that a value of k results in a single value of D.

CHAPTER 2

Frequency

Distributions

2.1   Data Description: Graphs

Repeated measurements yield data, which must be organized according to some principle. The data should be arranged in such a way that each observation can fall into one, and only one, category. A simple graphical method of presenting data is the pie chart, which is a circle divided into parts that represent categories.

EXAMPLE:

2000 Budget

38% came from individual income taxes
28% from social insurance receipts
13% from corporate income taxes
12% from borrowing
5% from excise taxes
4% other

This data can also be presented in the form of a bar chart or bar graph.

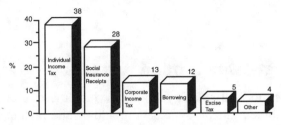

EXAMPLE:

The population of the United States for the years 1860 and 1960 is shown in the table below,

Year	1860	1870	1880	1890	1900	1910
Population in millions	31.4	39.8	50.2	62.9	76.0	92.0
Year	1920	1930	1940	1950	1960	
Population in millions	105.7	122.8	131.7	151.1	179.3	

in this graph,

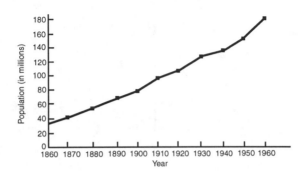

and in this bar chart.

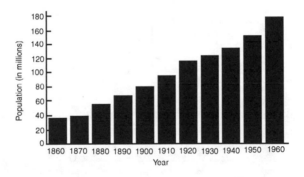

EXAMPLE:

A quadratic function is given by

$$y = x^2 + x - 2$$

We compute the values of y corresponding to various values of x.

x	-3	-2	-1	0	1	2	3
y	4	0	-2	-2	0	4	10

From this table, the points of the graph are obtained:

$$(-3, 4)\ (-2, 0)\ (-1, -2)\ (0, -2)\ (1, 0)\ (2, 4)\ (3, 10)$$

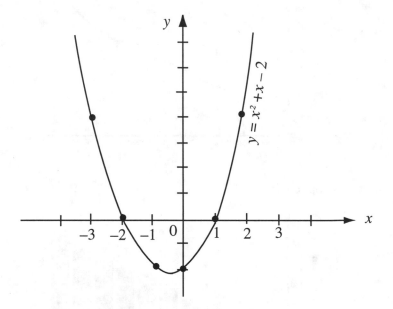

The curve shown is called a parabola. The general equation for a parabola is

$$y = ax^2 + bx + c, \quad a \neq 0$$

where *a, b,* and *c* are constants.

Problem Solving Example:

Twenty students are enrolled in the foreign language department, and their major fields are as follows: Spanish, Spanish, French, Italian, French, Spanish, German, German, Russian, Russian, French, German, German, German, Spanish, Russian, German, Italian, German, and Spanish.

(a) Make a frequency distribution table.

(b) Make a frequency bar graph.

(a) The frequency distribution table is constructed by writing down the major field and next to it the number of students.

Major Field	Number of Students
German	7
Russian	3
Spanish	5
French	3
Italian	2
Total	20

(b) A bar graph follows:

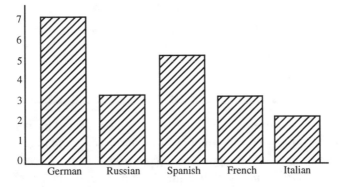

In the bar graph, the fields are listed and spaced evenly along the horizontal axis. Each specified field is represented by a rectangle, and all have the same width. The height of each, identified by a number on the vertical axis, corresponds to the frequency of that field.

A survey of 1,000 adults was conducted to determine what meal they preferred to have in a fast food restaurant. Forty percent preferred breakfast, 30% preferred lunch, 20% preferred dinner, and 10% preferred a snack. Display this information in a pie chart.

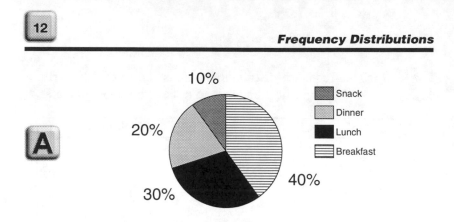

To determine the angular measurement of each section of the pie chart, convert the percentage to a decimal and multiply by 360°.

40% means a central angle of (.40) (360°) = 144°

30% means a central angle of (.30) (360°) = 108°

20% means a central angle of (.20) (360°) = 72°

10% means a central angle of (.10) (360°) = 36°

2.2 Class Intervals and Class Limits

A set of measurements that has not been organized numerically is called **raw data**. An arrangement of raw numerical data in descending or ascending order of magnitude is called an **array**. The difference between the largest and smallest numbers in a set of data is called the **range of the data**.

Problem Solving Example:

Find the range of the sample composed of the observations 3, 12, 15, 7, and 9.

The range is the difference between the largest and smallest observations. The largest observation is 15 and the smallest is 3. The range is 15 − 3 = 12.

EXAMPLE:

One hundred families were chosen at random, and their yearly income was recorded.

Income of 100 Families

Income in Thousands	Number of Families
10 – 14	3
15 – 19	12
20 – 24	19
25 – 29	20
30 – 34	23
35 – 39	18
40 – 44	5

Total   100

The range of the measurements is divided by the number of class intervals desired.

In this case, we have seven classes. The number of individuals belonging to each class is called the class frequency.

For example, the class frequency of the class 35 – 39 is 18. The list of classes, together with their class frequencies, is called a frequency table or frequency distribution.

The table is a frequency distribution of the income of 100 families. In the table, the labels 10 – 14 , 15 – 19, ..., 40 – 44 are called class intervals. For the interval 10 – 14, the numbers 14 and 10 are called class limits; 10 is the lower class limit, and 14 is the upper class limit.

In some cases, open class intervals are used; for example, "40,000 and over." In the table, income is recorded to the nearest thousand. Hence, the class interval 30 – 34 includes all incomes from 29,500 to 34,499.

The exact numbers 29,500 and 34,499 are called class boundaries. The smaller number is the lower class boundary, and the larger is the upper class boundary.

Class boundaries are often used to describe the classes. The difference between the lower and upper class boundaries is called the class width or class size. Usually, all class intervals of a frequency distribution have equal widths. In the table, the class width is 5.

The class mark (or class midpoint) is the midpoint of the class interval.

$$\text{Class Midpoint} = \frac{\text{Lower Class Limit} + \text{Upper Class Limit}}{2}$$

The class mark of the class interval 35 – 39 is

$$\frac{35 + 39}{2} = 37$$

For purposes of mathematical analysis, all data belonging to a given class interval are assumed to coincide with the class mark.

2.2.1 Guidelines for Constructing Class Intervals and Frequency Distributions

1. Find the range of the measurements, which is the difference between the largest and the smallest measurements.

2. Divide the range of the measurements by the approximate number of class intervals desired. The number of class intervals is usually between 5 and 20, depending on the data.

   Then, round the result to a convenient unit, which should be easy to work with. This unit is a common width for the class intervals.

3. The first class interval should contain the smallest measurement, and the last class interval should contain the largest measurement. All measurements should lie between class limits.

4. Determine the number of measurements that fall into each class interval; that is, find the class frequencies.

EXAMPLE:

The 25 measurements given below represent the sulphur level in the air for a sample of 25 days. The unit used is parts per million.

27	32	28	32	31
35	28	44	45	36
33	40	41	36	35
39	37	39	37	44
41	41	35	35	33

The lowest reading is 27 and the highest is 45.

Thus, the range is $45 - 27 = 18$. It will be convenient to use 5 class intervals: $\frac{18}{5} = 3.6$. We round 3.6 to 4. The width for the class intervals is 4. The class intervals are labeled as follows: $26 - 29$, $30 - 33$, $34 - 37$, $38 - 41$, and $42 - 45$.

Now, we can construct the frequency table and compute the class frequency for each class. The relative frequency of a class is defined as the frequency of the class divided by the total number of measurements. The table above shows the classes, frequencies, and relative frequencies of the data in the table below.

Class Interval	Frequency	Relative Frequency
26 – 29	3	0.12
30 – 33	5	0.20
34 – 37	8	0.32
38 – 41	6	0.24
42 – 45	3	0.12
Total	25	1.00

2.3 Frequency Histograms and Relative Frequency Histograms

2.3.1 Definitions of Histograms and Polygons

A frequency histogram (or histogram) is a set of rectangles placed in the coordinate system.

The vertical axis is labeled with the frequencies, and the horizontal axis is labeled with the class intervals. Over each class interval, a rectangle is drawn with a height such that the area of the rectangle is proportional to the class frequency.

Often, the height is numerically equal to the class frequency. Based on the results of the table on the bottom of the previous page, we obtain the histogram shown below.

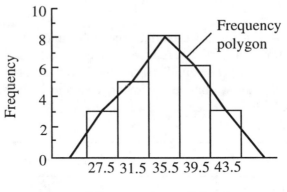

Measurements of Sulphur

A frequency polygon is obtained by connecting midpoints of the tops of the rectangles of the histogram.

The area bounded by the frequency polygon and the x-axis is equal to the sum of the areas of the rectangles in the histogram.

The relative frequency histogram is similar to the frequency histogram. In the histogram below, the vertical axis shows relative frequency. A rectangle is constructed over each class interval with a height equal to the class relative frequency. Based on the previous table (bottom of page 15), we obtain the relative frequency histogram shown.

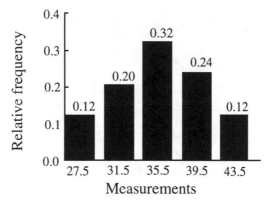

2.4   Cumulative Frequency Distributions

2.4.1   Definition of Cumulative Frequency

The total frequency of all values less than the upper class boundary of a given interval up to and including that interval is called the cumulative frequency.

When we know the class intervals and their corresponding frequencies, we can compute the cumulative frequency distribution.

Consider the class intervals and frequencies contained in the table on the bottom of page 15, from which we compute the cumulative frequency distribution. See the following table.

Parts of Sulphur	Number of Days
Less than 26	0
Less than 30	3
Less than 34	8
Less than 38	16
Less than 42	22
Less than 46	25

Using the coordinate system, we can present the cumulative frequency distribution graphically. Such a graph is called a cumulative frequency polygon or ogive. The ogive obtained from the table above is shown in the figure below.

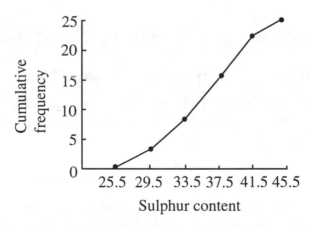

For a very large number of observations, it is possible to choose very small class intervals and still have a significant number of observations fall within each class.

The sides of the frequency polygon or relative frequency polygon get smaller as class intervals get smaller. Such polygons that closely approximate curves are called frequency curves or relative frequency curves. Usually, frequency curves can be obtained by increasing the number of class intervals, which requires a larger sample.

Problem Solving Examples:

Q The following data is a sample of the accounts receivable of a small merchandising firm.

37	42	44	47	46	50	48	52	90
54	56	55	53	58	59	60	62	92
60	61	62	63	67	64	64	68	
67	65	66	68	69	66	70	72	
73	75	74	72	71	76	81	80	
79	80	78	82	83	85	86	88	

Using a class interval of 5, i.e., 35 — 39,

(a) Make a frequency distribution table.
(b) Construct a histogram.
(c) Draw a frequency polygon.
(d) Make a cumulative frequency distribution.
(e) Construct a cumulative percentage ogive.

A

(a)

Interval	Class Boundaries	Class Tally	Interval Median	Frequency
35 – 39	34.5 – 39.5	/	37	1
40 – 44	39.5 – 44.5	//	42	2
45 – 49	44.5 – 49.5	///	47	3
50 – 54	49.5 – 54.5	////	52	4
55 – 59	54.5 – 59.5	////	57	4
60 – 64	59.5 – 64.5	╫╫ ///	62	8
65 – 69	64.5 – 69.5	╫╫ ///	67	8
70 – 74	69.5 – 74.5	╫╫ /	72	6
75 – 79	74.5 – 79.5	////	77	4
80 – 84	79.5 – 84.5	╫╫	82	5
85 – 89	84.5 – 89.5	///	87	3
90 – 94	89.5 – 94.5	//	92	2

We use the fractional class boundaries. One reason for this is that we cannot break up the horizontal axis of the histogram into only integral values. We must do something with the fractional parts. The usual thing to do is to assign all values to the closest integer. Hence, we use the class boundaries provided. The appropriate histogram follows.

(b)

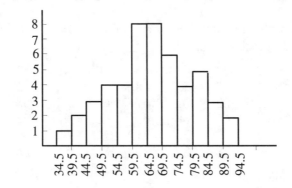

We now construct a frequency polygon as follows:

Plot points (x_i, f_i), where x_i is the interval median and f_i is the class frequency. Connect the points by successive line segments.

(c)

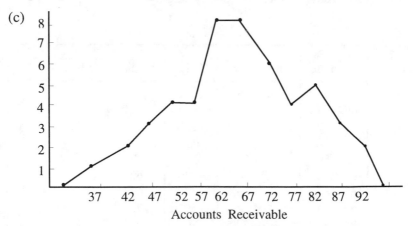

Accounts Receivable

(d)

Interval	Interval Median	Frequency (f_i)	Cumulative Frequency	Cumulative Percentage
35 – 39	37	1	1	2
40 – 44	42	2	3	6
45 – 49	47	3	6	12
50 – 54	52	4	10	20
55 – 59	57	4	14	28
60 – 64	62	8	22	44
65 – 69	67	8	30	60
70 – 74	72	6	36	72
75 – 79	77	4	40	80
80 – 84	82	5	45	90
85 – 89	87	3	48	96
90 – 94	92	2	50	100

The cumulative frequency is the number of values in all classes up to and including that class. It is obtained by addition. For example, the cumulative frequency for 65 – 69 is $1 + 2 + 3 + 4 + 4 + 8 + 8 = 30$. The cumulative percentage is the percent of all observed values found in that class or below. We can use the formula

$$\text{cumulative frequency} = \frac{\text{cumulative frequency}}{\text{total observations}} \times 100\%$$

For example, cum. per. (65 - 69) $= \dfrac{30}{50} \times 100\% = 60\%$

(e) We construct the cumulative percentage ogive by plotting points (x_i, f_i), where x_i is the interval median and f_i is the cumulative frequency. Finally, we connect the points with successive line segments.

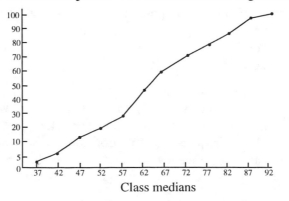

Class medians

Q In the following chart, make two additional columns and fill in the cumulative frequencies and cumulative percentages. Also, draw a histogram and a cumulative frequency diagram.

Relative Frequency Distribution of 100 Sixth-Grade Students and Their Weights

Class	Frequency (f_i)	Relative Frequency	Percentage Distribution
59 – 61	4	4/100	4
62 – 64	8	8/100	8
65 – 67	12	12/100	12
68 – 70	13	13/100	13
71 – 73	21	21/100	21
74 – 76	15	15/100	15
77 – 79	12	12/100	12
80 – 82	9	9/100	9
83 – 85	4	4/100	4
86 – 88	2	2/100	2
	100		100%

A The relative frequency of a class is found by dividing the class frequency by the total number of observations in the sample. The results, when multiplied by 100, form a percentage distribution. The class relative frequencies and the percentage distribution of the weights of 100 sixth-grade students are given in the above table.

The relative frequency of a class is the empirical probability that a random observation from the population will fall into that class. For example, the relative frequency of the class 59 – 61 in the table is 4/100, and, therefore, the empirical probability of a random observation falling in this interval is 4/100.

The table allows us to determine the percentage of the observations in a sample that lie in a particular class. When we want to know the percentage of observations that is above or below a specified interval, the cumulative frequency distribution can be used to our advantage. The cumulative frequency distribution is obtained by adding the frequencies in all classes less than or equal to the class with which we are concerned. To find the percentage in each class, just divide the frequency by the total number of observations and multiply by 100%. In this example,

$$\frac{x}{100} \times 100\% = x\%.$$

Now we can find the cumulative percentages by taking the cumulative frequencies.

$$\text{cum. percentage } = \frac{y \text{ cumulative frequency}}{\text{total observations}} \times 100\%$$

$$= \frac{y}{100} \times 100\% = y\%$$

Cumulative Frequency and Cumulative Percentage Distribution of the 100 Sixth-Grade Students and Their Weights

Class	Frequency (f_i)	Cumulative Frequency	Cumulative Percentage
59 – 61	4	4	4
62 – 64	8	12	12
65 – 67	12	24	24
68 – 70	13	37	37
71 – 73	21	58	58
74 – 76	15	73	73
77 – 79	12	85	85
80 – 82	9	94	94
83 – 85	4	98	98
86 – 88	2	100	100%
	Total 100		

The data in a frequency distribution may be represented graphically by a histogram. The histogram is constructed by marking off the class boundaries along a horizontal axis and drawing a rectangle to represent each class. The base of the rectangle corresponds to the class width, and the height to that class' frequency. See the accompanying histogram depicting the data on the table in the beginning of the problem. Note that the areas above the various classes are proportional to the frequency of those classes.

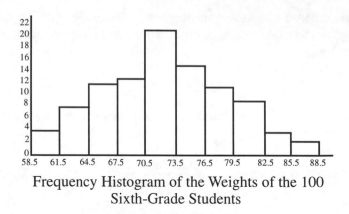

Frequency Histogram of the Weights of the 100
Sixth-Grade Students

Often a frequency polygon is used instead of a histogram. In constructing a frequency polygon, the points (x_i, f_i) are plotted on horizontal and vertical axes. The polygon is completed by adding a class mark with zero frequency to each end of the distribution and joining all the points with line segments. The frequency diagram for the data in this problem follows:

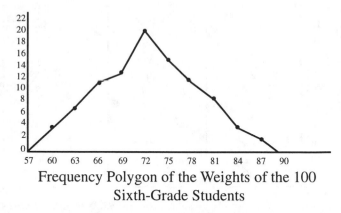

Frequency Polygon of the Weights of the 100
Sixth-Grade Students

A frequency polygon may also be constructed by connecting the midpoints of the bars in a frequency histogram by a series of line segments. The main advantage of the frequency polygon compared to the frequency histogram is that it indicates that the observations in the interval are not all the same. Also, when several sets of data are to be shown on the same graph, it is clearer to superimpose frequency polygons than histograms, especially if class boundaries coincide.

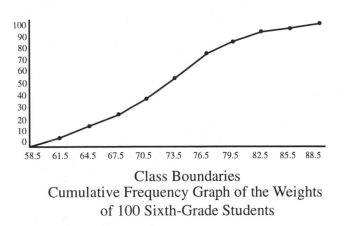

Class Boundaries
Cumulative Frequency Graph of the Weights
of 100 Sixth-Grade Students

It is often advantageous and desirable to make a graph showing the cumulative frequency within a sample. The data for such a graph depicting the cumulative frequency of the weights of the 100 sixth-grade students are found in column three of the table in the solution. The graph, called an ogive, is illustrated above. To avoid the confusion of less than or greater than, the class boundaries are plotted on the horizontal axis rather than the interval medians.

A cumulative frequency graph makes it easy to read such items as the percentage of students whose weights are less than or greater than a specified weight. If the cumulative percentage had been plotted, the graph would appear the same as above but would be called a percentage ogive.

2.5 Types of Frequency Curves

In applications we find that most of the frequency curves fall within one of the categories listed.

1. One of the most popular is the bell-shaped or symmetrical frequency curve.

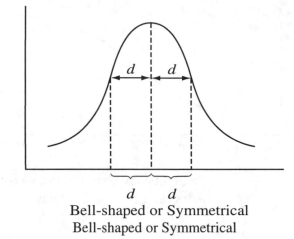

Bell-shaped or Symmetrical
Bell-shaped or Symmetrical

Note that observations equally distant from the maximum have the same frequency. The normal curve has a symmetrical frequency curve.

2. The U-shaped curve has maxima at both ends.

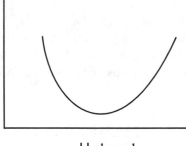

U-shaped

3. A curve can also be skewed to one side. Askew to the left is when the slope to the right of the maximum is steeper than the slope to the left. The opposite holds for the frequency curve skewed to the right.

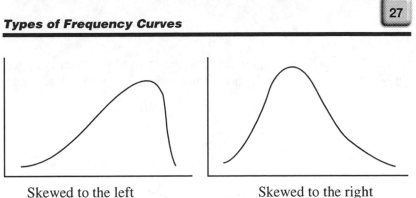

Skewed to the left
(negative skew)

Skewed to the right
(positive skew)

4. A J-shaped curve has a maximum at one end.

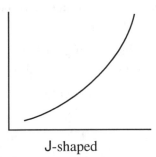

J-shaped

5. A multimodal (bimodal) frequency curve has two or more maxima.

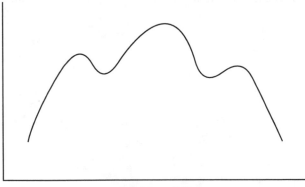

Multimodal

Q What are two ways to describe the form of a frequency distribution? How would the following distributions be described?

(a)   (b)

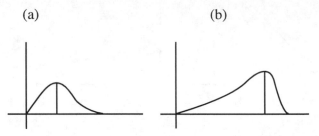

A The form of a frequency distribution can be described by its departure from symmetry or skewness and its degree of peakedness or kurtosis.

If the few extreme values are higher than most of the others, we say that the distribution is "positively skewed" or "skewed" to the right.

If the few extreme values are lower than most of the others, we say that the distribution is "negatively skewed" or "skewed" to the left.

(a) This distribution has extreme values in the upper half of the curve and is skewed to the right or positively skewed.

(b) The extreme values of this distribution are in the lower half of the curve. Thus, the distribution is negatively skewed or skewed to the left.

CHAPTER 3

Numerical Methods of

Describing Data

3.1   Introduction

Once we have a sufficient number of measurements, it is easy to find the frequency distribution. Graphic methods, however, are very often impractical or difficult to convey.

To remedy the situation, one can use a few numbers that describe the frequency distribution without drawing a real graph. Such numbers are called numerical descriptive measures, and each one describes a certain aspect of the frequency distribution. None of them yields the exact shape of the frequency distribution. Rather, they give us some notion of the general shape of the whole graph or parts of it.

For example, saying that somebody is 6' 4" and weighs 250 lbs. does not describe the person in detail, but it does give us the general idea of a stout man.

It is important to describe the center of the distribution of measurements as well as how the measurements behave about the center of the distribution. For that purpose, we define central tendency and variability. In practical applications, we deal with one of two essentially different situations:

1. The measurements are gathered about the whole population.

   Numerical descriptive measures for a population are called **parameters.**

2. The measurements are gathered about the sample. Numerical descriptive measures for a sample are called **statistics.**

If we only have statistics, we are not able to calculate the values of parameters. But, using statistics, we can make reasonable estimates of parameters that describe the whole population. The most popular mathematical means of describing frequency distribution is an average. An **average** is a value that is representative or typical of a set of measurements.

Usually, averages are called measures of central tendency. We will be using different kinds of averages, such as the arithmetic mean, the geometric mean, and the harmonic mean. A different average should be applied depending on the data, the purpose, and the required accuracy.

3.2   Notation and Definitions of Means

By

$$x_1, x_2, ..., x_n$$

we denote the measurements observed in a sample of size n. The letter i in x_i is called a subscript or index. It stands for any of the numbers $1, 2, ..., n$.

Throughout the book, we will be using the summation notation. The symbol

$$\sum_{i=1}^{n} x_i$$

denotes the sum of all x_i's, that is,

$$\sum_{i=1}^{n} x_i = x_1 + x_2 + ... + x_{n-1} + x_n$$

EXAMPLE:

$$\sum_{i=1}^{4} x_i y_i = x_1 y_1 + x_2 y_2 + x_3 y_3 + x_4 y_4$$

EXAMPLE:

Let a be a constant. Then

$$\sum_{k=1}^{n} a x_k = a x_1 + a x_2 + \ldots + a x_n$$

$$= a(x_1 + x_2 + \ldots + x_n)$$

$$= a \sum_{k=1}^{n} x_k$$

In general,

$$\sum a x_k = a \sum x_k$$

and

$$\sum (ax + by) = a \sum x + b \sum y$$

Often, when no confusion can arise, we write $\sum_{k} x_k$ instead of $\sum_{k=1}^{n} x_k$.

3.2.1 Definition of Arithmetic Mean

The arithmetic mean, or mean, of a set of measurements is the sum of the measurements divided by the total number of measurements.

The arithmetic mean of a set of numbers $x_1, x_2, \ldots, x_n$ is denoted by x (read "x bar").

$$\bar{x} = \frac{\sum_{i=1}^{n} x_1}{n} = \frac{x_1 + x_2 + \ldots + x_n}{n}$$

EXAMPLE:

The arithmetic mean of the numbers 3, 7, 1, 24, 11, and 32 is

$$\bar{x} = \frac{3 + 7 + 1 + 24 + 11 + 32}{6} = 13$$

EXAMPLE:

Let $f_1, f_2, \ldots, f_n$ be the frequencies of the numbers $x_1, x_2, \ldots, x_n$ (i.e., number x_i occurs f_i times). The arithmetic mean is

$$\bar{x} = \frac{f_1 x_1 + f_2 x_2 + \ldots + f_n x_n}{f_1 + f_2 + \ldots + f_n} = \frac{\sum_{i=1}^{n} f_i x_i}{\sum_{i=1}^{n} f_i}$$

$$= \frac{\sum fx}{\sum f}.$$

Note that the total frequency, that is, the total number of cases, is

$$\sum_{i=1}^{n} f_i.$$

EXAMPLE:

If the measurements 3, 7, 2, 8, 0, and 4 occur with frequencies 3, 2, 1, 5, 10, and 6, respectively, then the arithmetic mean is

$$\bar{x} = \frac{3 \times 3 + 7 \times 2 + 2 \times 1 + 8 \times 5 + 0 \times 10 + 4 \times 6}{3 + 2 + 1 + 5 + 10 + 6} \approx 3.3$$

Keep in mind that the arithmetic mean is strongly affected by extreme values.

EXAMPLE:

Consider four workers whose annual salaries are $2,500, $3,200, $3,700, and $48,000. The arithmetic mean of their salaries is

$$\frac{\$57,400}{4} = \$14,350$$

The figure $14,350 can hardly represent the typical annual salary of the four workers.

EXAMPLE:

The deviation d_i of x_i from its mean x is defined to be

$$d_i = x_i - \bar{x}$$

The sum of the deviations of $x_1, x_2, \ldots, x_n$ from their mean x is equal to zero. Indeed,

$$\sum_{i=1}^{n} d_1 = \sum_{i=1}^{n} (x_i - \bar{x}) = 0$$

Thus,

$$\sum_{i=1}^{n} (x_i - \bar{x}) = \sum_{i=1}^{n} x_i - n\bar{x} = \sum x_i - n \frac{\sum x_i}{n}$$

$$= \sum x_i - \sum x_i = 0.$$

EXAMPLE:

If $z_1 = x_1 + y_1, \ldots, z_n = x_n = y_n$, then $\bar{z} = \bar{x} + \bar{y}$. Indeed,

$$\bar{x} = \frac{\sum x}{n}, \bar{y} = \frac{\sum y}{n}, \text{ and } \bar{z} + \frac{\sum z}{n}.$$

We have

$$\bar{z} = \frac{\sum z}{n} = \frac{\sum (x+y)}{n} = \frac{\sum x}{n} + \frac{\sum y}{n} = \bar{x} + \bar{y}.$$

The arithmetic mean plays an important role in statistical inference.

We will be using different symbols for the sample mean and the population mean. The population mean is denoted by μ, and the sample mean is denoted by $\bar{x}$. The sample mean $\bar{x}$ will be used to make inferences about the corresponding population mean μ.

EXAMPLE:

Suppose a bank has 500 savings accounts. We pick a sample of 12 accounts. The balance on each account in dollars is

657	284	51
215	73	327
65	412	218
539	225	195

The sample mean $\bar{x}$ is

$$\bar{x} = \frac{\sum_{i=1}^{12} x_i}{12} = \$271.75$$

The average amount of money for the of 12 sampled accounts is $271.75. Using this information, we estimate the total amount of money in the bank to be

$$\$271.25 \times 500 = \$135,875.$$

Problem Solving Examples:

Q The following measurements were taken by an antique dealer as he weighed to the nearest pound his prized collection of anvils. The weights were 84, 92, 37, 50, 50, 84, 40, and 98. What was the mean weight of the anvils?

A The average or mean weight of the anvils is

$$\bar{x} = \frac{\text{sum of observations}}{\text{number of observations}}$$

$$= \frac{84 + 92 + 37 + 50 + 50 + 84 + 40 + 98}{8}$$

$$= \frac{535}{8} = 66.88 \cong 67 \text{ pounds}$$

An alternate way to compute the sample mean is to rearrange the terms in the numerator, grouping the numbers that are the same. Thus,

$$\bar{x} = \frac{(84 + 84) + (50 + 50) + 37 + 40 + 92 + 98}{8}$$

We see that we can express the mean in terms of the frequency of observations. The frequency of an observation is the number of times a number appears in a sample.

$$\bar{x} = \frac{2(84) + 2(50) + 37 + 40 + 92 + 98}{8}$$

The observations 84 and 50 appear in the sample twice, and thus each observation has frequency 2.

 The numbers 4, 2, 7, and 9 occur with frequencies 2, 3, 11, and 4, respectively. Find the arithmetic mean.

 To find the arithmetic mean, $\bar{x}$, multiply each different number by its associated frequency. Add these products, then divide by the total number of numbers.

$$\bar{x} = [(4)(2) + (2)(3) + (7)(11) + (9)(4)] \div 20$$
$$= (8 + 6 + 77 + 36) \div 20$$
$$= 127 \div 20 = 6.35$$

3.2.2 Definition of Geometric Mean

The geometric mean g (or G) of a set of n numbers

$$x_1, x_2, \ldots, x_n$$

is the nth root of the product of the numbers

$$g = \sqrt[n]{x_1 \times x_2 \times \ldots \times x_n}$$

EXAMPLE:

The geometric mean of the numbers 3, 9, and 27 is

$$g = \sqrt[3]{3 \times 9 \times 27} = 9$$

EXAMPLE:

Find the geometric mean of the numbers 3, 5, 7, 8, 10, 13, and 16. We begin with

$$g = \sqrt[7]{3 \times 5 \times 7 \times 8 \times 10 \times 13 \times 16} = \sqrt[7]{1,747,200}$$

Using common logarithms

$$\log g = \frac{1}{7} \log 1,747,200 = \frac{1}{7} \times 6.2423 = 0.8918$$

Hence,

$$g = 7.794$$

EXAMPLE:

The frequencies of measurements $x_1, x_2, \ldots, x_n$ are $f_1, f_2, \ldots, f_n$, respectively. The geometric mean of the measurements is

$$g = \sqrt[N]{\underbrace{x_1 \times x_1 \times \ldots \times x_1}_{f_1 \text{ times}} \times \underbrace{x_2 \times x_2 \times \ldots \times x_2}_{f_2 \text{ times}} \times \ldots \times \underbrace{x_n \times \ldots \times x_n}_{f_n \text{ times}}}$$

$$= \sqrt[N]{x_1^{f_1} \times x_2^{f_2} \times x_n^{f_n}} \text{ where } N = \sum_1^n f_i$$

If all the numbers are positive, then

$$\log g = \frac{1}{N} \log \left(x_1^{f_1} \times x_2^{f_2} \times \ldots \times x_n^{f_n} \right)$$

$$= \frac{1}{N} (f_1 \log x_1 + \ldots + f_n \log x_n)$$

$$= \frac{\sum f_1 \log x_i}{\sum f_i}$$

EXAMPLE:

The money deposited in an interest-bearing account increased from $1,000 to $5,000 in three years. What was the average percentage increase per year?

The increase is 500%. But, the average percentage increase is not $\frac{500\%}{3}$.

Denote the average percentage increase by q. Then

amount deposited   1,000

amount after first year $1{,}000 + q1{,}000 = 1{,}000(1 + q)$

amount after second year $1{,}000(1 + q) + 1{,}000(1 + q)q$

$$= 1{,}000(1 + q)^2$$

amount after third year $= 1{,}000(1 + q)^2 + 1{,}000(1 + q^2)q$

$$= 1{,}000(1 + q)^3$$

$$= 5{,}000$$

Hence,

$$(1 + q)^3 = 5$$

and

$$q = \sqrt[3]{5} - 1 \approx .71 = 71\%$$

In general, if the initial amount is A and the yearly interest is q, then after n years, the amount is M.

$$M = A(1 + q)^n$$

This equation is called the compound interest formula.

3.2.3 Definition of Weighted Arithmetic Mean

With the numbers $x_1, x_2, \ldots, x_n$ we associate weighting factors or weights, $w_1, w_2, \ldots, w_n$ depending on how significant each number is. The weighted arithmetic mean is defined by

$$\bar{x} = \frac{x_1 w_1 + x_2 w_2 + \ldots + x_n w_n}{w_1 + w_2 + \ldots + w_n} = \frac{\sum x_n w_n}{\sum w_n}$$

EXAMPLE:

The pilot has to pass three tests. The second test is weighted three times as much as the first, and the third test is weighted four times as much as the first. The pilot reached the score of 40 on the first test, 45 on the second test, and 60 on the third test.

The weighted mean is

$$\frac{40 \times 1 + 45 \times 3 + 60 \times 4}{8} = 51.875$$

EXAMPLE:

If f_1 numbers have mean m_1, f_2 numbers have mean m_2, ..., f_n numbers have mean m_n, then the mean of all numbers is

$$\bar{x} = \frac{f_1 m_1 + f_2 m_2 + ... + f_n m_n}{f_1 + f_2 + ... + f_n}$$

Problem Solving Example:

Q A student takes two quizzes, one midterm, and one final exam in a statistics course. The midterm counts three times as much as a quiz, and the final exam counts five times as much as a quiz. If the quiz scores were 70 and 80, the midterm score was 65, and the final exam score was 85, what was the weighted average?

A $x = [(70)(1) + (80)(1) + (65)(3) + (85)(5)] \div 10$

$= (70 + 80 + 195 + 425) \div 10$

$= 770 \div 10 = 77$

3.2.4 Definition of Harmonic Mean

The harmonic mean h of numbers $x_1, x_2, ..., x_n$ is the reciprocal of the arithmetic mean of the reciprocals of the numbers:

$$h = \frac{1}{\frac{1}{n}\sum_{i=1}^{n}\frac{1}{x_i}} = \frac{n}{\sum\frac{1}{x_i}}$$

EXAMPLE:

The harmonic mean of the numbers 2, 4, and 9 is

$$h = \frac{3}{\frac{1}{2}+\frac{1}{4}+\frac{1}{9}} = 3.48$$

EXAMPLE:

During four successive years, the prices of gasoline were 70, 75, 78, and 95 cents per gallon. Find the average cost of gasoline over the four-year period. There are two methods of computing this average.

Method 1

Suppose the car owner used 100 gallons each year. Then

$$\text{Average Cost} = \frac{\text{total cost}}{\text{total number of gallons}}$$

$$= \frac{0.7\times100 + 0.75\times100 + 0.78\times100 + 0.95\times100}{400}$$

$$= 0.795 \ \$/\text{gal}$$

Method 2

Suppose the car owner spends $100 on gasoline each year. Then

$$\text{Average Cost} = \frac{\text{total cost}}{\text{total number of gallons}}$$

$$= \frac{400}{509.66} = 0.785 \ \$/gal$$

Both averages are correct. Depending on assumed conditions, we obtain different answers.

Let g be the geometric mean of a set of positive numbers $x_1, x_2, ..., x_n$, and let $\bar{x}$ be the arithmetic mean and h the harmonic mean. Then

$$h \leq g \leq \bar{x}$$

The equality holds only if all the numbers $x_1, x_2, ..., x_n$ are identical.

Problem Solving Example:

Q A motor car traveled 3 consecutive miles, the first mile at $x_1 = 35$ miles per hour (mph), the second at $x_2 = 48$ mph, and the third at $x_3 = 40$ mph. Find the average speed of the car in miles per hour.

A Distance = Rate × Time. Therefore, Time = $\dfrac{\text{Distance}}{\text{Rate}}$.

For the first mile, Time = $\dfrac{1 \text{ mile}}{35 \text{ miles/hour}} = \dfrac{1}{35}$ hour.

For the second mile, Time = $\dfrac{1 \text{ mile}}{48 \text{ miles/hour}} = \dfrac{1}{48}$ hour.

For the third mile, Time = $\dfrac{1 \text{ mile}}{40 \text{ miles/hour}} = \dfrac{1}{40}$ hour.

Total time = $T_1 + T_2 + T_3 = \dfrac{1}{35} + \dfrac{1}{48} + \dfrac{1}{40}$.

Converting to decimals, $\text{Time}_{tot} = .0286 + .0208 + .025$

$$= .0744 \text{ hours.}$$

The average speed can be computed by the following formula:

Average Speed $= \dfrac{\text{Total distance}}{\text{Total time}} = \dfrac{3 \text{ miles}}{.0744 \text{ hours}} = 40.32$ mph.

Average speed is an example of a harmonic mean. The harmonic is

$$h = \dfrac{3}{\dfrac{1}{T_1} + \dfrac{1}{T_2} + \dfrac{1}{T_3}}.$$

3.2.5 Definition of Root Mean Square

The root mean square, or quadratic mean, of the numbers $x_1, x_2, \ldots,$ x_n is denoted by $\sqrt{\overline{x^2}}$ and defined as

$$\sqrt{\overline{x^2}} = \sqrt{\dfrac{\displaystyle\sum_{i=1}^{n} x_i^2}{n}}$$

EXAMPLE:

The quadratic mean of the numbers 2, 3, 5, and 7 is

$$\sqrt{\overline{x^2}} = \sqrt{\dfrac{2^2 + 3^2 + 5^2 + 7^2}{4}} = 4.66$$

EXAMPLE:

The quadratic mean of two positive numbers, a and b, is not smaller than their geometric mean.

$$\sqrt{ab} \leq \sqrt{\dfrac{a^b + b^2}{2}}$$

Indeed, since

$$0 \leq (a - b)^2 = a^2 - 2ab + b^2$$

we obtain

$$ab \leq \dfrac{a^2 + b^2}{2}$$

and

$$\sqrt{ab} \leq \sqrt{\dfrac{a^b + b^2}{2}}$$

All means can also be computed for the grouped data, that is, when data are presented in a frequency distribution. Then, all values within a given class interval are considered to be equal to the class mark, or midpoint, of the interval.

3.3 Measures of Central Tendency

3.3.1 Definition of the Mode

The mode of a set of numbers is that value which occurs most often (with the highest frequency).

Observe that the mode may not exist. Also, if the mode exists, it may not be unique. For example, for the numbers 1, 1, 2, and 2, the mode is not unique.

EXAMPLE:

The set of numbers 2, 2, 4, 7, 9, 9, 13, 13, 13, 26, and 29 has mode 13.

The set of numbers that has two or more modes is called **bimodal.**

For grouped data – data presented in the form of a frequency table – we do not know the actual measurements, only how many measurements fall into each interval. In such a case, the mode is the midpoint of the class interval with the highest frequency.

Note that the mode can also measure popularity. In this sense, we can determine the most popular model of car or the most popular actor.

EXAMPLE:

One can compute the mode from a histogram or frequency distribution.

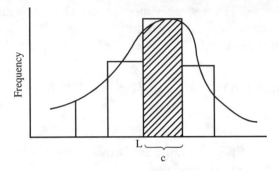

The shaded area indicates the modal class, that is, the class containing the mode.

$$Mode = L + c\left[\frac{\Delta_1}{\Delta_1\Delta_2}\right]$$

where L is the lower class boundary of the modal class

      c is the size of the modal class interval

      Δ_1 is the excess of the modal frequency over the frequency of the next lower class

      Δ_2 is the excess of the modal frequency over the frequency of the next higher class

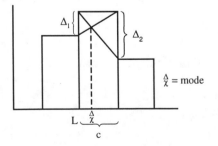

Problem Solving Examples:

 Find the mode of the sample 14, 19, 16, 21, 18, 19, 24, 15, and 19.

 The mode is another measure of central tendency in a data set. It is the observation or observations that occur with the greatest frequency. The number 19 is observed three times in this sample, and no other observation appears as frequently. The mode of this sample is therefore 19.

 Find the mode or modes of the sample 6, 7, 7, 3, 8, 5, 3, and 9.

 In this sample the numbers 7 and 3 both appear twice. There are no other observations that appear as frequently as these two. Therefore, 3 and 7 are the modes of this sample. The sample is called "bimodal."

 Find the mode of the sample 14, 16, 21, 19, 18, 24, and 17.

 In this sample all the numbers occur with the same frequency. There is no single number that is observed more frequently than any other. Thus, there is no mode or all observations are modes. The mode is not a useful concept here.

3.3.2 Definition of Median

The median of a set of numbers is defined as the middle value when the numbers are arranged in order of magnitude.

Usually, the median is used to measure the midpoint of a large set of numbers. For example, we can talk about the median age of people getting married. Here, the median reflects the central value of the data for a large set of measurements. For small sets of numbers, we use the following conventions:

— For an odd number of measurements, the median is the middle value.

— For an even number of measurements, the median is the average of the two middle values.

In both cases, the numbers have to be arranged in order of magnitude.

EXAMPLE:

The scores of a test are 78, 79, 83, 83, 87, 92, and 95. Hence, the median is 83.

EXAMPLE:

The median of the set of numbers 21, 25, 29, 33, 44, and 47 is $\frac{29+33}{2} = 31$.

It is more difficult to compute the median for grouped data. The exact value of the measurements is not known; hence, we know only that the median is located in a particular class interval. The problem is where to place the median within this interval.

For grouped data, the median obtained by interpolation is given by

$$\text{Median} = L + \frac{c}{f_{\text{median}}} \left(\frac{n}{2} - \left(\sum f \right) \text{cum} \right)$$

where	L	=	the lower class limit of the interval that contains the median
c	=	the size of the median class interval
f_{median}	=	the frequency of the median class
n	=	the total frequency
$(\Sigma f)_{\text{cum}}$	=	the sum of frequencies (cumulative frequency) for all classes before the median class

Problem Solving Examples:

 Find the median of the sample 34, 29, 26, 37, and 31.

 The median, a measure of central tendency, is the middle number. The number of observations that lie above the median is the same as the number of observations that lie below it.

Arranged in order we have 26, 29, 31, 34, and 37. The number of observations is odd, and thus the median is 31. Note that there are two numbers in the sample above 31 and two below 31.

 Find the median of the sample 34, 29, 26, 37, 31, and 34.

 The sample arranged in order is 26, 29, 31, 34, 34, and 37. The number of observations is even and thus the median, or middle number, is chosen halfway between the third and fourth numbers. In this case, the median is

$$\frac{31+34}{2} = 32.5$$

EXAMPLE:

The weight of 50 men is depicted in the table below in the form of frequency distribution.

Weight	Frequency
115 – 121	2
122 – 128	3
129 – 135	13
136 – 142	15
143 – 149	9
150 – 156	5
157 – 163	3
Total	50

Class 136 – 142 has the highest frequency.

The mode is the midpoint of the class interval with the highest frequency.

$$\text{Mode} = \frac{135.5 + 142.5}{2} = 139$$

We can also use the formula

$$\text{Mode} = L + c\left(\frac{\Delta_1}{\Delta_1 + \Delta_2}\right)$$

where

L = 135.5

c = 7

Δ^1 = 2 (15 – 13 = 2)

Δ^2 = 6 (15 – 9 = 6)

$$\text{Mode} = 135.5 + 7 \times \frac{2}{2+6} = 137.25$$

The median is located in class 136 – 142.

We have

$$\text{Median} = L + \frac{c}{f_{\text{median}}}\left[\frac{n}{2} - \left[\sum f\right]_{\text{cum}}\right]$$

where

L = 135.5

c = 7

f_{median} = 15

n = 50

$(\Sigma f)_{\text{cum}}$ = 2 + 3 + 13 = 18

Hence,

$$\text{Median} = 135.5 + \frac{7}{15}\left[\frac{50}{2} - 18\right] = 138.77$$

To compute the arithmetic mean for grouped data, we compute midpoint x_i of each of the intervals and use the formula

$$\bar{x} = \frac{\sum\limits_{i=1}^{n} f_i x_i}{\sum\limits_{i=1}^{n} f_i}$$

We have

$$\bar{x} = \frac{118 \times 2 + 125 \times 3 + 132 \times 13 + 139 \times 15 + 146 \times 9 + 153 \times 5 + 160 \times 3}{50}$$

$$= 139.42.$$

For symmetrical curves, the mean, mode, and median all coincide.

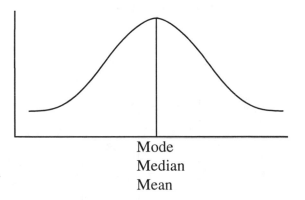

Mode
Median
Mean

For skewed distributions, we have the following.

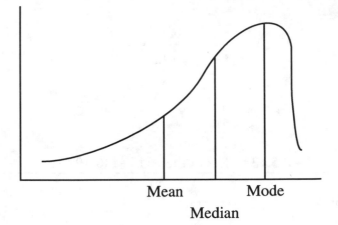

The distribution is skewed to the left.

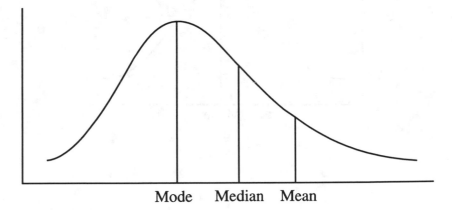

The distribution is skewed to the right.

Problem Solving Examples:

Class Boundaries	Class Weights	Frequencies
58.5 – 61.5	60	4
61.5 – 64.5	63	8
64.5 – 67.5	66	12
67.5 – 70.5	69	13
70.5 – 73.5	72	21
73.5 – 76.5	75	15
76.5 – 79.5	78	12
79.5 – 82.5	81	9
82.5 – 85.5	84	4
85.5 – 88.5	87	2

 Find the median weight from the previous table.

 There are 100 observations in the sample. The median will be the 50th observation. When using an even-numbered sample of grouped data, the convention is to call the $\frac{n}{2}$th observation the median. There are 37 observations in the first four intervals, and the first five intervals contain 58 observations. The 50th observation is in the fifth class interval.

We use the technique of linear interpolation to estimate the position of the 50th observation within the class interval.

The width of the fifth class is three, and there are 21 observations in the class. To interpolate we imagine that each observation takes up $\frac{3}{21}$ units of the interval. There are 37 observations in the first four intervals, and thus the 13th observation in the fifth class will be the

median. This 13th observation will be approximately $13\left(\dfrac{3}{21}\right)$ units from

the lower boundary of the fifth class interval. The median is thus the

lower boundary of the fifth class plus $13\left(\dfrac{3}{21}\right)$ or

$$\text{median} = 70.5 + \frac{13}{7} = 72.36.$$

A sample of drivers involved in motor vehicle accidents was categorized by age. The results appear as:

Age	Number of Accidents
16 – 25	28
26 – 35	13
36 – 45	12
46 – 55	8
56 – 65	19
66 – 75	20

What is the value of the median?

We seek the $\dfrac{100}{2}$ = 50th number, which appears in the third class (36 – 45).

The total number of accidents is 100. The median is the $\dfrac{100}{2}$ = 50th number when the numbers are arranged in ascending order. (In this case, we have intervals of numbers instead of just numbers.) The two intervals 16 – 25 and 26 – 35 consist of 41 count. We need nine numbers from the interval 36 – 45. Use the lower boundary of this interval 36 – 45, which is 35.5, and add $\dfrac{9}{12}$ of the width of the interval (10). Then

$$35.5 + \frac{9}{12}(10) = 43$$

Quiz: Introduction – Numerical Methods of Describing Data

1. A numerical sample is grouped into the following classes:

Class	Limits
I	0 – 8
II	9 – 17
III	18 – 26
IV	27 – 35
V	36 – 44
VI	45 – 53

    The width, or real size, of these classes is

    (A) 8.0.   (D) 9.5.

    (B) 8.5.   (E) 10.0.

    (C) 9.0.

2. Given the ogive of a frequency distribution, which classes have equal frequencies?

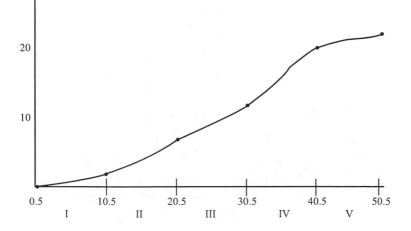

(A) I and V

(D) III and V

(B) II and III

(E) None of the above.

(C) III and IV

3.   The class limits of a numerical sample are given as follows:

Class	Limits
I	10 – 19
II	20 – 29
III	30 – 39
IV	40 – 49
V	50 – 59
VI	60 – 69

What are the class boundaries of Class IV?

(A) 40, 49

(D) 39.9, 49.1

(B) 39.95, 49.05

(E) 39.5, 49.5

(C) 39, 50

The following histogram shows the distribution of admissions test scores obtained by applicants to a junior college.

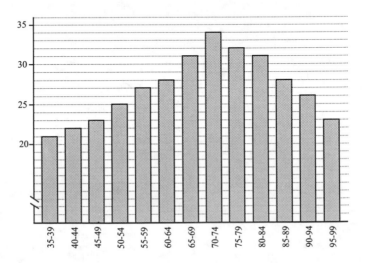

4.   How many candidates scored less than 55 points?

(A) 118   (D) 27

(B) 91   (E) 25

(C) 63

5.   A class has four quizzes worth 30% (w1) of their grade, three tests worth 40% (w2), and a final worth 30% (w3). If these are Jenny's scores, what will her grade-point average be?

quizzes: 67, 75, 85, and 72

tests: 85, 95, and 80

final: 83

(A) 87.12%   (D) 80.25%

(B) 77.12%   (E) 82%

(C) 81.47%

6.   The diagram shows the graph of a frequency distribution.

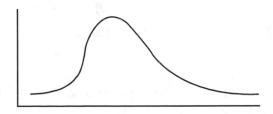

From left to right, what is the order of the measures of central location?

(A) Median, mode, mean

(B) Mode, median, mean

(C) Mean, median, mode

(D) Mean, mode, median

(E) Can't tell from the diagram.

7. Given the frequency table:

Class	Class Boundaries	Frequency
I	0.5 – 10.5	6
II	10.5 – 20.5	8
III	20.5 – 30.5	12
IV	30.5 – 40.5	7
V	40.5 – 50.5	5
VI	50.5 – 60.5	2

What is the median for this grouped data?

(A) 20.5   (D) 30

(B) 25.5   (E) 30.5

(C) 27.5

8. In a study of people's work habits, 16 federal employees are chosen at random, and the number of days they worked during one month is determined. The average number of days worked per person is 20. If 12 of the people worked 19 days during the month, how many days did the other four people work?

(A) 20   (D) 23

(B) 21   (E) 24

(C) 22

9. Given the following table of grouped data, what is the mode of this data given as a single value?

Class	Class Limits	Frequency
I	50 – 54	12
II	55 – 59	17
III	60 – 64	9
IV	65 – 69	4
V	70 – 74	2
VI	75 – 79	1

The mode is

(A) 22.   (D) 57.

(B) 23.   (E) 59.5.

(C) 64.5.

10. A sample of six data is arranged in ascending order. The lowest value is 8 and the range of the data is 10. Which of the following statements *must* be true?

(A) The median equals the third number.

(B) The mode is the value of the sixth number.

(C) The mean is greater than 10.

(D) The mode is 10.

(E) The median equals the mean of the third and fourth numbers.

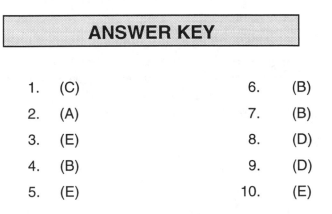

ANSWER KEY

1.	(C)	6.	(B)
2.	(A)	7.	(B)
3.	(E)	8.	(D)
4.	(B)	9.	(D)
5.	(E)	10.	(E)

CHAPTER 4

Measures of Variability

4.1 Range and Percentiles

The degree to which numerical data tend to spread about an average value is called the **variation** or **dispersion** of the data. We shall define various measures of dispersion.

The simplest measure of data variation is the range.

4.1.1 Definition of Range

The **range** of a set of numbers is defined to be the difference between the largest and the smallest number of the set. For grouped data, the range is the difference between the upper limit of the last interval and the lower limit of the first interval.

EXAMPLE:

The range of the numbers 3, 6, 21, 24, and 38 is $38 - 3 = 35$.

4.1.2 Definition of Percentiles

The nth percentile of a set of numbers arranged in order of magnitude is that value which has $n\%$ of the numbers below it and $(100 - n)\%$ above it.

EXAMPLE:

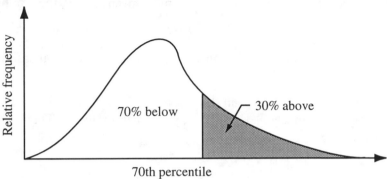

70th percentile
The 70th percentile of a set of numbers

Percentiles are often used to describe the results of achievement tests. For example, someone graduates in the top 10% of the class. Frequently used percentiles are the 25th, 50th, and 75th percentiles, which are called the lower quartile, the middle quartile (median), and the upper quartile, respectively.

4.1.3 Definition of Interquartile Range

The interquartile range, abbreviated IQR, of a set of numbers is the difference between the upper and lower quartiles.

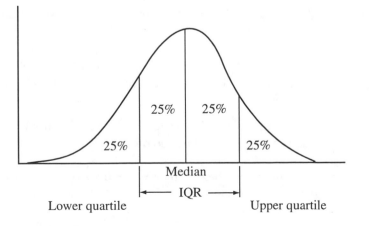

Now, we shall introduce an important concept of deviation.

The deviation of a number x from its mean $\bar{x}$ is defined to be

$$x - \bar{x}$$

Using deviations, we can construct many different measures of variability.

Observe that the mean deviation for any set of measurements is always zero. Indeed, let $x_1, x_2, \dots, x_n$ be measurements. Their mean is given by

$$\bar{x} = \frac{\sum x_i}{n}$$

The deviations are $x_1 - \bar{x}, x_2 - \bar{x}, \dots, x_n - \bar{x}$, and their mean is equal to

$$\frac{\sum_{i=1}^{n} (x_i - \bar{x})}{n} = \frac{\sum x_i}{n} - \bar{x} = 0$$

Problem Solving Examples:

 Find the range of the sample composed of the observations 33, 53, 35, 37, and 49.

 The range is a measure of the dispersion of the sample and is defined to be the difference between the largest and smallest observations.

In our sample, the largest observation is 53 and the smallest is 33. The difference is $53 - 33 = 20$, and the range is 20.

The range is not a very satisfactory measure of dispersion as it involves only two of the observations in the sample.

In a sample of data, the 75th percentile is the number 23. If the interquartile range is 10, what number represents the 25th percentile?

The interquartile range = the difference between the 75th percentile and the 25th percentile. If x = 25th percentile, we have $23 - x = 10$, so $x = 13$.

4.2 Measures of Dispersion

4.2.1 Definition of Average Deviation

The average deviation of a set of n numbers $x_1, x_2, \ldots, x_n$ is defined by

$$\text{Average Deviation} = \frac{\sum_{i=1}^{n} |x_i - \bar{x}|}{n} = \overline{|x - \bar{x}|}$$

where $\bar{x}$ is the arithmetic mean of the numbers $x_1, x_2, \ldots, x_n$.

EXAMPLE:

We find the average deviation of the numbers 3, 5, 6, 8, 13, and 21 by

$$\bar{x} = \frac{3 + 5 + 6 + 8 + 13 + 21}{6} = 9.33$$

$$\text{Average Deviation} = \frac{6.33 + 4.33 + 3.33 + 1.33 + 3.67 + 11.67}{6}$$

$$= 5.11$$

If the frequencies of the numbers $x_1, x_2, \ldots, x_n$ are $f_1, f_2, \ldots, f_n$, respectively, then the average deviation becomes

$$\text{Average Deviation} = \frac{\sum_i f_i |x_i - \bar{x}|}{\sum f_i}$$

We will be using this formula for grouped data where x_i's represent class marks and f_i represents class frequencies.

The sum $\displaystyle\sum_{i=1}^{n}|x_i - a|$ is the maximum when a is the median.

Problem Solving Example:

Compute the average deviation for the following sample representing the age at which men in a Chataqua bowling club scored their first game over 175.

$$29, 36, 42, 48, 49, 56, 59, 62, 64, 65$$

The average deviation is the average absolute deviation from the mean. The average deviation is an alternate measure of variation in a sample. It is defined as

$$\text{A.D.} = \frac{\sum |X_i - \overline{X}|}{n},$$

where X_i are the individual observations, $\overline{X}$ the sample mean, and n the number of observations.

$|X_i - \overline{X}|$ = the absolute value of the difference between the ith observation and $\overline{X}$.

$$\overline{X} = \frac{\sum X_i}{n} = \frac{29 + 36 + 42 + 48 + 49 + 56 + 59 + 62 + 64 + 65}{10}$$

$$= \frac{510}{10} = 51$$

$$\sum |X_i - \overline{X}| = |29 - 51| + |36 - 51| + |42 - 51| + |48 - 51| + |49 - 51| +$$

$$|56 - 51| + |59 - 51| + |62 - 51| + |64 - 51| + |65 - 51|$$

$$= 22 + 15 + 9 + 3 + 2 + 5 + 8 + 11 + 13 + 14 = 102$$

Thus, $\text{A.D.} = \dfrac{102}{n} = \dfrac{102}{10} = 10.2.$

4.2.2 Definition of Standard Deviation

The standard deviation of a set $x_1, x_2, ..., x_n$ of n numbers is defined by

$$s = \sqrt{\frac{\sum_{i=1}^{n}(x_i - \bar{x})^2}{n}} = \sqrt{\overline{(x - \bar{x})^2}}$$

The sample standard deviation is denoted by s, while the corresponding population standard deviation is denoted by σ.

For grouped data, we use the modified formula for standard deviation. Let the frequencies of the numbers $x_1, x_2, ..., x_n$ be $f_1, f_2, ..., f_n$, respectively. Then,

$$s = \sqrt{\frac{\sum f_i (x_i - \bar{x})^2}{\sum f_i}} = \sqrt{\frac{\sum f (x - \bar{x})^2}{\sum f}}$$

Often, in the definition of the standard deviation, the denominator is not n but $n - 1$. For large values of n, the difference between the two definitions is negligible.

4.2.3 Definition of Variance

The variance of a set of measurements is defined as the square of the standard deviation. Thus,

$$s^2 = \frac{\sum_{i=1}^{n}(x_i - \bar{x})^2}{n-1}$$

or

$$s^2 = \frac{\sum_{i=1}^{n} f_i(x_i - \bar{x})^2}{\sum_{i=1}^{n} f_i}$$

Usually, the variance of the sample is denoted by s^2, and the corresponding population variance is denoted by σ^2.

EXAMPLE:

A simple manual task was given to six children, and the time each child took to complete the task was measured. Results are shown in the table.

x_i	$x_i - \bar{x}$	$(x_i - \bar{x})^2$
12	2.5	6.25
9	−0.5	0.25
11	1.5	2.25
6	−3.5	12.25
10	0.5	0.25
9	−0.5	0.25
Total 57	0	21.5

For this sample, we shall find the standard deviation and variance.

The average $\bar{x}$ is 9.5.

$$\bar{x} = 9.5$$

The standard deviation is

$$s^2 = \sqrt{\frac{21.5}{5}} = 2.07$$

and the variance is

$$s^2 = 4.3$$

Problem Solving Examples:

 A couple has six children whose ages are 6, 8, 10, 12, 14, and 16. Find the variance in ages.

 The variance in ages is a measure of the spread or dispersion of ages about the sample mean.

To compute the variance, we first calculate the sample mean.

$$\bar{X} = \frac{\sum X_i}{n} = \frac{\text{sum of observations}}{\text{number of observations}}$$

$$= \frac{6+8+10+12+14+16}{6} = \frac{66}{6} = 11$$

The variance is defined to be

$$s^2 = \frac{\sum_{i=1}^{n}(X_i - \bar{X})^2}{n-1}$$

$$= \frac{(6-11)^2 + (8-11)^2 + (10-11)^2 + (12-11)^2 + (14-11)^2 + (16-11)^2}{5}$$

$$= \frac{25+9+1+1+9+25}{5} = \frac{70}{5} = 14$$

Q In a particular sample with 20 observations, the numbers 4, 6, 7, and 9 occur with frequencies 7, 3, 8, and 2, respectively. Find the value of s^2.

A $\overline{X} = \sum_{i=1}^{4} X_i f_i \div n,$ where X_i = each individual number, f_i = its respective frequency, and n = total frequency.

s^2 = variance of a sample = $\dfrac{\sum_{i=1}^{4}(X_i - \overline{X})^2 \times f_i}{n-1}$ (note: $n - 1$ in the denominator for a sample)

So, $\overline{X} = [(4)(7) + (6)(3) + (7)(8) + (9)(2)] \div 19 = 120 \div 19 \approx 6$

Then, $s^2 = [7(4 - 6)^2 + 3(6 - 6)^2 + 8(7 - 6)^2 + 2(9 - 6)^2] \div 19$

$= 54 \div 19 \approx 2.84$

4.3 Simplified Methods for Computing the Standard Deviation and Variance

By definition,

$$s^2 = \frac{\sum_{i=1}^{n}(x_i - \overline{x})^2}{n-1}$$

Hence, through the process of deviation

$$s^2 = \frac{n\left(\sum x^2\right) - \left(\sum x\right)^2}{n(n-1)}$$

and

$$s = \sqrt{\frac{n\left(\sum x^2\right) - \left(\sum x\right)^2}{n(n-1)}}$$

We find

$$s^2 = \frac{n\left(\sum f \times x^2\right) - \left(\sum f \times x\right)^2}{n(n-1)}$$

EXAMPLE:

Consider the example in Section 4.2.3. We shall use the formula

$$s^2 = \frac{n\left(\sum x^2\right) - \left(\sum x\right)^2}{n(n-1)}$$

to compute the value of s^2.

$$s^2 = \frac{6(563) - (57)^2}{6(5)} = 4.3$$

and

$$s^2 = \sqrt{4.3} = 2.07$$

Remember that $\bar{x}^2$ denotes the mean of the squares of all values of x and $\bar{x}^2$ denotes the square of the mean of all x's.

Let D be an arbitrary constant and

$$d_i = x_i - D$$

be the deviations of x_i from D. Then

$$x_i = d_i + D$$

and

$$\bar{x} = \bar{d} + D$$

Hence,

$$s^2 = \overline{(x - \bar{x})^2} = \overline{(d - \bar{d})^2} = \overline{d^2 - 2\bar{d}d + \bar{d}^2}$$

$$= \overline{d^2} - \bar{d}^2 = \frac{\sum fd^2}{\sum f} - \left(\frac{\sum fd}{\sum f}\right)^2$$

EXAMPLE:

We shall find the standard deviation for the data shown below. The height of 100 students was measured and recorded.

Height (inches)	Class Mark x	x^2	Frequency f	fx^2	fx
60 – 62	61	3,721	7	26,047	427
63 – 65	64	4,096	21	86,016	1,344
66 – 68	67	4,489	37	166,093	2,479
69 – 71	70	4,900	26	127,400	1,820
72 – 74	73	5,329	9	47,961	657

$$\sum f = 100 \qquad \sum fx^2 = 453,517 \qquad \sum fx = 6,727$$

We apply the formula

$$s = \sqrt{\frac{n\left(\sum f \times x^2\right) - \left(\sum f \times x\right)^2}{n(n-1)}}$$

Hence

$$s = \sqrt{\frac{100(453,517) - (6,727)^2}{100(99)}} = 3.165$$

Using the empirical rule, we can interpret the standard deviation of a set of measurements.

4.3.1 Empirical Rule

If a set of measurements has a bell-shaped histogram, then

1. the interval $x \pm s$ contains approximately 68% of the measurements.

2. $\bar{x} \pm 2s$ contains approximately 95% of the measurements.

3. $\bar{x} \pm 3s$ contains approximately all the measurements.

Note that from the empirical rule we can find the approximate value of the sample standard deviation s. Approximately 95% of all the measurements are located in the interval $\bar{x} \pm 2s$. The length of this interval is $4s$. Hence, the range of the measurements is equal to approximately $4s$.

$$\text{Approximate Value of } s = \frac{\text{range}}{4}$$

For normal (bell-shaped) distributions we have

Empirical Rule

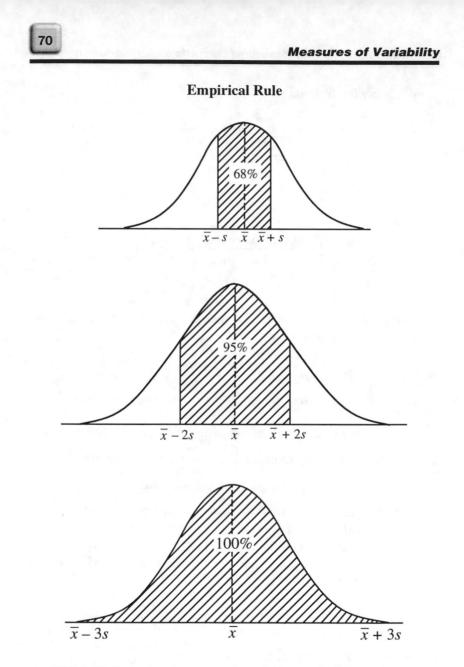

We shall discuss some properties of the standard deviation.
In the definition of the standard deviation

$$s = \sqrt{\frac{\sum_{i=1}^{n}(x_i - \bar{x})^2}{n-1}}$$

we can replace $\bar{x}$ by any other average y that is not the arithmetic mean

$$s = \sqrt{\frac{\sum(x_i - y)^2}{n-1}}$$

Note that of all such standard deviations, the one with $y = \bar{x}$ is the minimum. Indeed

$$\frac{\sum(x - y)^2}{n} = \frac{\sum(x^2 - 2xy + y^2)}{n-1}$$

$$= \frac{\sum x^2 - 2y\sum x + ny^2}{n-1}$$

$$= y^2 - 2y\frac{\sum x}{n-1} + \frac{\sum x^2}{n-1}$$

Equation $y^2 + ay + b$, where a and b are constants, has a minimum value if, and only if, $y = -\frac{1}{2}a$. Hence,

$$y = \frac{\sum x}{n-1} = \bar{x}$$

Consider two sets consisting of N_1 and N_2 measurements or two frequency distributions with total frequencies N_1 and N_2. The variances are $s_1{}^2$ and $s_2{}^2$, respectively.

$$s^2 = \frac{N_1 s_1^2 + N_2 s_2^2}{N_1 + N_2}$$

If both sets have the same mean $\bar{x}$, then the combined (or pooled) variance of both sets or both frequency distributions is given by

$$s^2 = \frac{N_1 s_1 + N_2 s_2{}^2}{N_1 + N_2}$$

Problem Solving Examples:

Q Using the chart below, find the variance and standard deviation of the weights of the sixth-grade students.

Class Boundaries	Class Weights	Frequencies
58.5 – 61.5	60	4
61.5 – 64.5	63	8
64.5 – 67.5	66	12
67.5 – 70.5	69	13
70.5 – 73.5	72	21
73.5 – 76.5	75	15
76.5 – 79.5	78	12
79.5 – 82.5	81	9
82.5 – 85.5	84	4
85.5 – 88.5	87	2

A The variance of grouped data is similar to the variance for ungrouped data. In both cases, the variance is an average of the squared deviation from $\bar{X}$.

We have seen that for grouped data

The variance for grouped data is

$$s^2 = \frac{\sum f_i(X_i - \overline{X})^2}{\sum f_i},$$

where f_i is the frequency or number of observations in the ith class, X_i is the midpoint of the ith class, and $\overline{X}$ is the sample mean of the grouped data.

This variance can be computed more conveniently using the computational formula derived below,

$$s^2 = \frac{\sum f_i(X_i - \overline{X})^2}{\sum f_i} = \frac{\sum f_i(X_i^2 - 2X_i\overline{X} + \overline{X}^2)}{\sum f_i}$$

$$= \frac{\sum f_i X_i^2 - 2X\sum f_i X_i + \overline{X}^2 \sum f_i}{\sum f_i}$$

$$\left(\text{because } \sum f_i X_i = \overline{X}^2 \sum f_i\right)$$

$$= \frac{\sum f_i X_i^2 - 2\left(\sum f_i\right)\overline{X}^2 + \overline{X}^2 \sum f_i}{\sum f_i}$$

$$= \frac{\sum f_i X_i^2}{\sum f_i} = \overline{X}^2.$$

We have computed $\overline{X}$ to be 72.45.

This table is convenient in computing $\sum f_i X_1^2$.

Class mark X_i	Frequency f_i	X_i^2	$f_i X_i^2$
60	4	3,600	14,400
63	8	3,969	31,752
66	12	4,356	52,272
69	13	4,761	61,893
72	21	5,184	108,864
75	15	5,625	84,375
78	12	6,084	73,008
81	9	6,561	59,049
84	4	7,056	28,224
87	2	7,569	15,138

$$\sum_{i=1}^{10} f_i X_1^2 = 528,975$$

$$s^2 = \frac{\sum f_i X_1^2}{\sum f_i} - (\overline{X})^2$$

The standard deviation is

$$= \frac{100(528,975) - (7,245)^2}{100(99)}$$

$$= \sqrt{41.16} = 6.42$$

$$s = \sqrt{s^2} = \sqrt{40.75} = 6.38.$$

A history test was taken by 51 students. The scores ranged from 50 to 95 and were classified into eight classes with a width of six units. The resulting frequency distribution appears below. Find s^2 by applying the definition for s^2. Then find s.

Class i	Class Mark X_i'	Frequency f_i	$X_i' f_i$
1	51	2	102
2	57	3	171
3	63	5	315
4	69	8	552
5	75	10	750
6	81	12	972
7	87	10	870
8	93	1	93

$$51 \qquad 3{,}825 = \sum_{i=1}^{8} X_i' f_i$$

 We wish to find s^2, the measure of dispersion of the observations about the sample mean for this classified data.

In the table we are given:

X_i' = the midpoint (class mark) of the ith class

f_i = the number of observations in the ith class

n = the total number of observations = $\displaystyle\sum_{i=1}^{n} f_i$

By definition, the sample variance for classified data is

$$s^2 = \frac{\displaystyle\sum_{i=1}^{K} (X_i' - \overline{X})^2 f_i}{n-1}$$

where K is the number of classes. First find $\overline{X}$.

$$\overline{X} = \frac{\sum\limits_{i=1}^{K} X_i' f_i}{n} = \frac{3,825}{51} = 75.$$

The computations used in finding s^2 are displayed in the following table.

Class	X_i'	f_i	$X_i' - \overline{X}$	$(X_i' - \overline{X})^2$	$(X_i' - X)^2 f_i$
1	51	2	− 24	576	1,152
2	57	3	− 18	324	972
3	63	5	− 12	144	720
4	69	8	− 6	36	288
5	75	10	0	0	0
6	81	12	6	36	432
7	87	10	12	144	1,440
8	93	1	18	324	324

$$\sum_{i=1}^{8} f_i = 50, \qquad \sum_{i=1}^{8} (X_i' - \overline{X}) f_i = 5,328$$

$$s^2 = \frac{1}{n}\sum_{i=1}^{8}(X_i' - \overline{X})^2 f_i = \frac{1}{51}(5,328) = 106.56$$

and the standard deviation,

$$s = \sqrt{s^2} = 10.32$$

 Two samples have the same mean. The first sample has 10 observations with a variance of 3. The second sample has 15 observations with a variance of 4. What is the combined variance?

 The combined (or pooled) variance = the sum of the product of each number of observations and its respective variance, divided by the total number of observations.

The combined (or pooled) variance $= [(10)(3) + (15)(4)] \div 25 = 3.6$

$$[30 + 60] \div 25 = 3.6$$

$$90 \div 25 = 3.6$$

4.4 Coding Methods

We use the coding methods to simplify calculations. Data are generally coded by one or both of the following operations:

1. addition (or subtraction) of a constant A to (or from) each measurement.

2. multiplication (or division) of each measurement by a constant.

We shall describe the results of coding. Let x_1, x_2, ..., x_n be n measurements with the arithmetic mean $\bar{x}$ and sample standard deviation s.

1. If a constant A was subtracted from each measurement, the mean and the standard deviation are given by

$$\bar{x}_A = \bar{x} - A$$

$$s_A = s$$

2. If each measurement was multiplied by a positive constant k, the mean and the standard deviation are given by

$$\bar{x}_k = k\bar{x}$$

$$s_k = ks$$

Next, we shall describe the coding procedure for grouped data.

The frequency distribution is given with the class intervals of equal size c. Each class mark x_i is coded into a corresponding value y_i by

$$x_i = A + cy_i$$

where A and c are constants. The standard deviation is

$$s = \sqrt{\overline{(x - \bar{x})^2}}$$

but

$$x = A + cy$$

and

$$\bar{x} = A + c\bar{y}$$
$$x - \bar{x} = c(y - \bar{y}).$$

Thus,

$$s^2 = \overline{(x - \bar{x})^2} = \overline{c^2(y - \bar{y})^2}$$
$$= \overline{c^2(y^2 - 2y\bar{y} + \bar{y}^2)} = c^2(\overline{y^2} - 2\bar{y}^2 + \bar{y}^2)$$
$$= c^2(\overline{y^2} - \bar{y}^2)$$

and

$$s = c\sqrt{\overline{y^2} - \bar{y}^2} = c\sqrt{\frac{\sum fy^2}{\sum f} - \left(\frac{\sum fy}{\sum f}\right)^2}$$

4.4.1 Charlier's Check

We shall describe Charlier's check using an example.

EXAMPLE:

The results of the I.Q. (intelligence quotient) test of 150 students are shown in the table. The data are grouped, and only class marks are recorded.

Class Mark	95	100	105	110	115	120
Frequency	3	5	9	14	17	21
Class Mark	125	130	135	140	145	150
Frequency	25	19	17	11	7	2

Using the coding method, we find the mean and the standard deviation.

Class mark 125 has the highest frequency. We denote

$$A = 125$$

The class intervals are of equal size $k = 5$.

Now, using

$$x_i = A + ky_i$$

and making substitutions,

$$130 = 125 + 5\,(1) \quad \text{and}$$

$$110 = 125 + 5\,(-3).$$

For each x_i, we compute y_i. The mean x can be calculated from the equation

$$\bar{x} = A + k\bar{y}$$

where

$$\bar{y} = \frac{\sum f_i y_i}{\sum f_i}$$

The standard deviation is given by

$$s = k\sqrt{\overline{y^2} - \bar{y}^2} = k\sqrt{\frac{\sum fy^2}{\sum f} - \left(\frac{\sum fy}{\sum f}\right)^2}$$

where

$$x_i = A + ky_i.$$

The results for each interval are listed in the table below.

	x_i	y_i	f_i	f_iy_i	$f_iy_i^2$
	95	−6	3	−18	108
	100	−5	5	−25	125
	105	−4	9	−36	144
	110	−3	14	−42	126
	115	−2	17	−34	68
	120	−1	21	−21	21
$A =$	125	0	25	0	0
	130	1	19	19	19
	135	2	17	34	68
	140	3	11	33	99
	145	4	7	28	112
	150	5	2	10	50

$$\sum f_i = 150 \quad \sum f_iy_i = -52 \quad \sum f_iy_i^2 = 940$$

The mean is

$$\bar{x} = A + k\bar{y} = A + k\frac{\sum f_iy_i}{\sum f_i}$$

$$= 125 + 5 \times \frac{-52}{150} = 123.27$$

and the standard deviation is

$$s = k \sqrt{\frac{\sum f_i y_i^2}{\sum f_i} - \left(\frac{\sum f_i y_i}{\sum f_i}\right)^2}$$

$$= 5 \sqrt{\frac{940}{150} - \left(\frac{-52}{150}\right)^2} = 12.4$$

Charlier's check was the identity

$$\sum f_i(y_i + 1) = \sum f_i y_i + \sum f_i$$

to verify the mean and the identity

$$\sum f_i(y_i + 1)^2 = \sum f_i y_i^2 + 2\sum f_i y_i + \sum f_i$$

to verify the value of the standard deviation. The values of $f_i(y_i + 1)$ and of $f_i(y_i + 1)^2$ are given in the table on the next page.

Also, the sums $\sum f_i(y_i + 1)$ and $\sum f_i(y_i + 1)^2$ are computed.

We have

$$\sum f_i y_i = -52$$
$$\sum f_i = 150$$

and

$$\sum f_i(y_i + 1) = 98$$

Indeed

$$-52 + 150 = 98$$

$y_i + 1$	f_i	$f_i(y_i + 1)$	$f_i(y_i + 1)^2$
−5	3	−15	75
−4	5	−20	80
−3	9	−27	81
−2	14	−28	56
−1	17	−17	17
0	21	0	0
1	25	25	25
2	19	38	76
3	17	51	153
4	11	44	176
5	7	35	175
6	2	12	72

$$\sum f_i = 150, \sum f_i(y_i + 1) = 98, \sum f_i(y_i + 1)^2 = 986$$

We have

$$\sum f(y+1)^2 = 986$$
$$\sum fy^2 = 940$$
$$2\sum fy = -104$$
$$\sum f = 150$$

Indeed

$$986 = 940 + 2(-52) + 150.$$

4.4.2 Sheppard's Correction

The value of the standard deviation s and of variance s^2 depends on how the data are grouped into classes. The error that occurs is called the grouping error. To compensate for the grouping error, the adjustment is introduced

$$\text{Corrected Variance} = \text{Calculated Variance} - \frac{k^2}{12}$$

where k is the class interval size. $\dfrac{k^2}{12}$ is called Sheppard's correction.

When Sheppard's correction should be applied depends on the situation.

For the normal distribution, the mean deviation is equal to 0.7979 times the standard deviation, and the semi-interquartile range is equal to 0.6745 times the standard deviation. For the normal distribution, we have the formulas

$$\text{Mean Deviation} = \frac{4}{5} \times \text{Standard Deviation}$$

$$\text{Semi-interquartile Range} = \frac{2}{3} \times \text{Standard Deviation}$$

4.4.3 Standardized Variable

We define the new variable

$$y = \frac{x - \bar{x}}{s}$$

which is called a standardized variable. It measures the deviation from the mean in units of the standard deviation. This variable is dimensionless, i.e., it is independent of the units used.

4.4.4 Coefficient of Variation

We shall call the actual dispersion or variation the absolute dispersion. The absolute dispersion is determined from the standard deviation or any other measure of dispersion.

The same value of dispersion can have an entirely different meaning in different situations. This fact is taken into account in the relative dispersion

$$\text{Relative Dispersion} = \frac{\text{Absolute Dispersion}}{\text{Average}}$$

When the absolute dispersion is the standard deviation, we obtain

$$\frac{s}{\overline{x}} = \text{Coefficient of Variation}$$

Observe that the coefficient of variation loses its meaning when x is close to zero.

Problem Solving Examples:

 The radii of five different brands of softballs (in inches) are 2.03, 1.98, 2.24, 2.17, and 2.08. Find the coefficient of variation.

 The coefficient of variation is defined as

$$V = \frac{s}{\overline{X}} = \frac{.105}{2.1} = .05.$$

Sometimes we want to compare sets of data that are measured differently. Suppose we have a sample of executives with a mean age of 51 and a standard deviation of 11.74 years. Suppose also we know their average IQ is 125 with a standard deviation of 20 points. How can we compare deviations? We use the coefficient of variation:

$$V_{age} = \frac{s}{\overline{X}} = \frac{11.74}{51} = .23; \quad V_{IQ} = \frac{s}{\overline{X}} = \frac{20}{125} = .16.$$

We now know that there is more variation with respect to age.

In our example, $V = \frac{s}{\overline{X}} = \frac{.105}{2.10} = .05.$

In a sample of data, the mean is 30, and the standard deviation is 2. If 27 is a member of this sample, what is its standardized value?

Standardized value $z = (X - \overline{X}) \div s$, where $\overline{X}$ is the mean and s is the standard deviation. So,

$$Z = \frac{(27 - 30)}{2}$$
$$= -\frac{3}{2} = -1.5$$

CHAPTER 5

Moments and Parameters
of Distributions

5.1 Moments

The set of numbers x_1, x_2, ..., x_n is given. Their sth moment is defined by

$$\overline{x^s} = \frac{x_1^s + x_2^s + \ldots + x_n^s}{n} = \frac{\sum\limits_{i=1}^{n} x_i^s}{n}$$

For $s = 1$ the first moment is the arithmetic mean $\overline{x}$.

The sth moment about the mean $\overline{x}$ is defined as

$$m_s = \frac{\sum\limits_{i=1}^{n}(x_i - \overline{x})^s}{n} = \overline{(x - \overline{x})^s}$$

EXAMPLE:

The numbers are 2, 3, 5, and 9. The first moment for this set of numbers is

$$\bar{x} = \frac{2+3+5+9}{4} = 4.75$$

which is also the arithmetic mean.

The second moment is

$$\overline{x^2} = \frac{2^2+3^2+5^2+9^2}{4} = 29.75$$

The third moment is

$$\overline{x^3} = \frac{2^3+3^3+5^3+9^3}{4} = 222.25$$

The fourth moment is

$$\overline{x^4} = \frac{2^4+3^4+5^4+9^4}{4} = 1{,}820.75$$

EXAMPLE:

The measurements are 1, 3, 5, and 15. The mean is $\bar{x} = 6$.

The first moment about the mean is

$$m_1 = \frac{(1-6)+(3-6)+(5-6)+(15-6)}{4} = 0$$

Note that the first moment about the mean is always equal to zero. Indeed,

$$m_1 = \frac{\sum(x-\bar{x})}{n} = \frac{\sum x}{n} - \bar{x} = 0$$

The second moment about the mean is

$$m_2 = \overline{(x - \bar{x})^2}$$
$$= \frac{(1-6)^2 + (3-6)^2 + (5-6)^2 + (15-6)^2}{4} = 29$$

Observe that m_2 is the variance o^{-2}.

The third moment about the mean is

$$m_3 = \overline{(x - \bar{x})^3}$$
$$= \frac{(1-6)^3 + (3-6)^3 + (5-6)^3 + (15-6)^3}{4} = 144$$

We define the sth moment about any number A (called origin) as

$$m_s' = \frac{\sum\limits_{i=1}^{n}(x_i - A)^s}{n} = \overline{(x - A)^s}$$

We have

$$m_s' = \frac{\sum d_i^s}{n}$$

where $d_i = x_i - A$ is the deviation of x_i from A.

Similarly, the moments for grouped data can be computed. Let x_1, x_2, ..., x_n be numbers which occur with frequencies f_1, f_2, ..., f_n, respectively.

The sth moment is defined by

$$\overline{x^s} = \frac{f_1 x_1^s + \ldots + f_n x_n^s}{f_1 + \ldots + f_n} = \frac{\sum f_i x_i^s}{\sum f_i}$$

The sth moment about the mean is defined by

$$m_s = \frac{\sum\limits_{i=1}^{n} f_i (x_i - \overline{x})^s}{\sum\limits_{i=1}^{n} f_i}$$

The sth moment about the origin A is defined as

$$m_s' = \frac{\sum\limits_{i=1}^{n} f_i (x_i - A)^s}{\sum\limits_{i=1}^{n} f_i}$$

There are some relations between moments. We shall list a few of them.

1. $$m_2 = m_2' - m_1'^2$$

$$m_2 = \overline{(x - \overline{x})^2} \text{ and } d = x - A.$$

Then

$$x = d + A \text{ and } \overline{x} = \overline{d} - A$$
$$x - \overline{x} = d - \overline{d}$$

We have

$$m_2 = \overline{(d - \overline{d})^2} = \overline{d^2 - 2\overline{d}d + \overline{d}^2}$$

$$= \overline{d^2} - 2\overline{d}\overline{d} + \overline{d}^2 = \overline{d^2} - \overline{d}^2$$

$$= m_2' - m_1'^2$$

2. $$m_3 = m_3' - 3m_1'm_2 + 2m_1'^3$$

$$m_3 = \overline{(x - \overline{x})^3} = \overline{(d - \overline{d})^3} = \overline{d^3 - 3d^2\overline{d} + 3d\overline{d}^2 - \overline{d}^3}$$

$$= \overline{d^3} - 3\overline{d^2}\overline{d} + 3\overline{d}^3 - \overline{d}^3 = \overline{d^3} - 3\overline{d}\,\overline{d^2} + 2\overline{d}^3$$

$$= m_3' - 3m_1'm_2 + 2m_1'^3$$

3. Similarly, we prove

$$m_4 = m_4' - 4m_1'm_3' + 6m_1'^2m_2' - 3m_1'^4$$

and formulas for higher moments.

Any moment m_k can be expressed in terms of moments $m_1{'}, m_2{'}, \ldots, m_k{'}$.

5.1.1   Computation of Moments for Grouped Data

The coding method can be applied when the data x_i can be expressed in the form

$$x_i = A + cy_i$$

Equation

$$m_s' = \frac{\sum f_i(x_i - A)^s}{\sum f_i}$$

becomes

$$m'_s = c^s \frac{\sum f_i y_i^s}{\sum f_i} = c^s \overline{y^s}$$

Which in turn enables us to compute moments m_s.

To verify the results, we can use Charlier's check. The moments are calculated by the coding method and the identities are applied.

$$\sum f(y+1) = \sum fy + \sum f$$

$$\sum f(y+1)^2 = \sum fy^2 + 2\sum fy + \sum f$$

$$\sum f(y+1)^3 = \sum fy^3 + 3\sum fy^2 + 3\sum fy + \sum f$$

$$\sum f(y+1)^4 = \sum fy^4 + 4\sum fy^3 + 6\sum fy^2 + 4\sum fy + \sum f$$

Sometimes it is more convenient to use dimensionless units.

We define the dimensionless moments about the mean

$$P_r = \frac{m_r}{s_r} = \frac{m_r}{\left(\sqrt{m_2}\right)^r} = \frac{m_r}{\sqrt{m_2^r}}$$

where $s = \sqrt{m_2}$ is the standard deviation.

Problem Solving Examples:

A sample consists of 3, 4, 7, 12, and 14. What is the third moment about the mean?

Since the mean is 8,

$$m_3 = \left[(3-8)^3 + (4-8)^3 + (7-8)^3 + (12-8)^3 + (14-8)^3\right] \div 5$$

$$= (-125 - 64 - 1 + 64 + 216) \div 5$$

$$= 90 \div 5 = 18$$

Q What is the value of m_4 (4th moment about the mean) in the previous problem?

A $m_4 = [(3 - 8)^4 + (4 - 8)^4 + (7 - 8)^4 + (12 - 8)^4 + (14 - 8)^4] \div 5$
$= (625 + 256 + 1 + 256 + 1{,}296) \div 5$

$= 2{,}434 \div 5 = 486.8$

5.2 Coefficients of Skewness: Kurtosis

Distributions can be symmetric or asymmetric. For example, the normal distribution is symmetric. Among the asymmetric distributions, some can be "more" asymmetric than others, and the degree of asymmetry is measured by skewness. Consider, for example, a distribution skewed to the right.

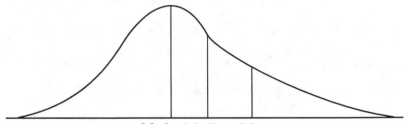

Mode   Median   Mean

The distribution skewed to the right has positive skewness. The asymmetry is measured by the difference.

5.2.1 Mean–Mode

To make it dimensionless, we can divide mean–mode by a measure of dispersion, such as the standard deviation. Thus,

$$\text{Skewness} = \frac{\text{Mean} - \text{Mode}}{\text{Standard Deviation}} = \frac{\bar{x} - \text{mode}}{s} (*)$$

For a distribution skewed to the right, mean > mode and skewness is positive. The skewness of a distribution skewed to the left is negative. Using the empirical relation

$$\text{Mean} - \text{Mode} = 3 (\text{Mean} - \text{Median})$$

we can write

$$\text{Skewness} = \frac{3(\text{Mean} - \text{Mode})}{\text{Standard Deviation}} (**)$$

Equation (*) is called Pearson's first coefficient of skewness and equation (**) is called Pearson's second coefficient of skewness.

There are also other measures of skewness, like

$$\text{Moment coefficient of skewness} = a_3 = \frac{m_3}{s^3} = \frac{m_3}{\sqrt{m_2^3}}$$

Also, $a_3^2 = \frac{m_3^2}{m_2^3}$ is used as a measure of skewness.

Skewness measures the degree of asymmetry. Kurtosis measures the shape of the peak of a distribution. Usually, this is measured relative to a normal distribution. A distribution can have one of three kinds of peaks. They are:

1. Leptokurtic, where a distribution has a relatively high peak.

2. Mesokurtic, where the peak is neither very high nor very low; a good example is the peak of the normal distribution.

3. Platykurtic, where the distribution is flat and the peak is low and not sharply outlined.

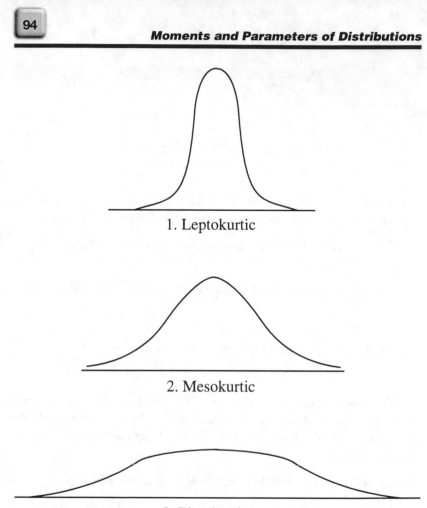

1. Leptokurtic

2. Mesokurtic

3. Platykurtic

Like skewness, kurtosis is described and measured in many ways. One measure of kurtosis is moment coefficient of kurtosis.

$$\text{Moment coefficient of kurtosis} = a_4 = \frac{m_4}{s^4} = \frac{m_4}{m_2^2}$$

For the normal distribution $a_4 = 3$.

Problem Solving Example:

Q What is the relative measure of skewness for the data listed below? This data represents the waist measurements of six randomly selected chocolate rabbits.

3 inches, 2 inches, 3.7 inches, 5 inches, 2.7 inches, 3 inches

 The relative measure of symmetry is defined to be $a_3 = \dfrac{m_3}{s^3}$,

where s^3 is the standard deviation cubed and m_3 is the third moment about the mean.

The third moment about the mean is defined as:

$$m_3 = \frac{\sum(x_i - \bar{x})^3}{n}.$$

We have encountered other examples of moments. The first moment about the mean is

$$m_1 = \frac{\sum(x_i - \bar{x})^1}{n}.$$

We can see that this moment has only one value.

$$m_1 = \frac{\sum(x_i - \bar{x})^1}{n} = \frac{\sum x_1}{n} = \frac{n\bar{x}}{n} = \frac{\sum x_i}{n} - \bar{x}$$

but $\bar{x} = \dfrac{\sum x_i}{n}$; thus, $m_1 = 0$.

The second moment about the mean is $\dfrac{\sum(x_i - \bar{x})^2}{n}$ or the sample variance.

The fourth moment is defined as

$$m_4 = \frac{\sum(x_i - \bar{x})^4}{n}$$

The measure of symmetry has the following interpretation, if $a_3 = \dfrac{m_3}{s^3}$ is equal to zero, the distribution is symmetrical. If $a_3 < 0$, then the distribution is negatively skewed. If $a_3 > 0$, the distribution is positively skewed.

To calculate the measure of symmetry, we use the table below.

x_i	$\bar{x}$	$(x_i - \bar{x})$	$(x_i - \bar{x})^2$	$(x_i - \bar{x})^3$
3	3.23	$-.23$	.053	$-.012$
2	3.23	-1.23	1.51	-1.86
3.7	3.23	.47	.22	.103
5	3.23	1.77	3.13	5.54
2.7	3.23	$-.53$	.28	$-.148$
3	3.23	$-.23$	.053	$-.012$

$$\sum(x_i - \bar{x})^2 = 5.246 \qquad \sum(x_i - \bar{x})^3 = 3.611$$

$$s^2 = \frac{\sum(x_i - \bar{x})^2}{n-1} = \frac{5.246}{5} = 1.05$$

$$s = \sqrt{s^2} = 1.02$$

$$s^3 = 1.15$$

$$m_3 = \frac{\sum(x_i - \bar{x})^3}{n} = \frac{3.611}{6} = .6018$$

and $\qquad a_3 = \dfrac{m_3}{s^3} = \dfrac{.6018}{1.15} = .5233.$

The distribution of the chocolate rabbits' waist measurements is skewed to the right, or positively skewed.

CHAPTER 6

Probability Theory: Basic Definitions and Theorems

6.1 Classical Definition of Probability

Let e denote an event that can happen in k ways out of a total of n ways. All n ways are equally likely. The probability of occurrence of the event e is defined as

$$p = pr\{e\} = \frac{k}{n}$$

The probability of occurrence of e is called its success. The probability of failure (non-occurrence) of the event is denoted by q.

$$q = pr\{\text{not } e\} = \frac{n-k}{n} = 1 - \frac{k}{n} = 1 - p$$

Hence, $p + q = 1$. The event "not e" is denoted by $\tilde{e}$ or $\sim e$.

EXAMPLE:

A toss of a coin will produce one of two possible outcomes heads or tails. Let e be the event that tails will turn up in a single toss of a coin.

Then

$$p = \frac{1}{1+1} = \frac{1}{2}.$$

EXAMPLE:

We define the event e to be number 5 or 6 turning up in a single toss of a die. There are six equally likely outcomes of a single toss of a die.

$$\{1, 2, 3, 4, 5, 6\}$$

Thus, $n = 6$. The event e can occur in two ways:

$$p = pr\{e\} = \frac{2}{6} = \frac{1}{3}.$$

Probability of failure of e is

$$q = pr\{\sim e\} = 1 - \frac{1}{3} = \frac{2}{3}.$$

For any event e,

$$0 \leq pr\{e\} \leq 1$$

If the event cannot occur, its probability is 0. If the event must occur, its probability is 1.

Next, we define the odds. Let p be the probability that an event will occur. The odds in favor of its occurrence are $p : q$ and the odds against it are $q : p$.

EXAMPLE:

We determine the probability that at least one tail appears in two tosses of a coin. Let h denote heads and t tails. The possible outcomes of two tosses are

$$(h, h), (h, t), (t, h), (t, t)$$

Three cases are favorable. Thus, p (success) $= \frac{3}{4}$ and p (failure) $= \frac{1}{4}$.

The odds in favor of at least one tail are $\frac{3}{4} : \frac{1}{4} = 3 : 1$.

EXAMPLE:

The event *e* is that the sum 8 appears in a single toss of a pair of dice. There are $6 \times 6 = 36$ outcomes:

$$(1, 1), (2, 1), (3, 1), \dots , (6, 6).$$

The sum 8 appears in five cases:

$$(2, 6), (6, 2), (3, 5), (5, 3), (4, 4).$$

Then

$$p\{e\} = \frac{5}{36}$$

The concept of probability is based on the concept of random experiment. A random experiment is an experiment with more than one possible outcome, conducted such that it is not known in advance which outcome will occur. The set of possible outcomes is denoted by a capital letter, say, *X*. Usually, each outcome is either a number (a toss of a die) or something to which a number can be assigned (heads = 1, tails = 0 for a toss of a coin).

For some experiments, the number of possible outcomes is infinite.

Problem Solving Examples:

What is the probability of throwing a 6 with a single die?

The die may land in any of six ways: 1, 2, 3, 4, 5, or 6. The probability of throwing a 6,

$$P(6) = \frac{\text{number of ways to get a } 6}{\text{number of ways the die may land}}$$

Thus, $P(6) = \frac{1}{6}$.

What is the probability of making a 7 in one throw of a pair of dice?

 There are 6 × 6 = 36 ways that two dice can be thrown, as shown below.

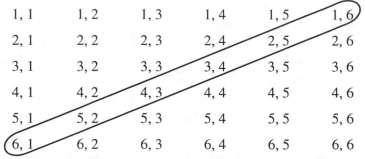

1, 1	1, 2	1, 3	1, 4	1, 5	1, 6
2, 1	2, 2	2, 3	2, 4	2, 5	2, 6
3, 1	3, 2	3, 3	3, 4	3, 5	3, 6
4, 1	4, 2	4, 3	4, 4	4, 5	4, 6
5, 1	5, 2	5, 3	5, 4	5, 5	5, 6
6, 1	6, 2	6, 3	6, 4	6, 5	6, 6

The number of possible ways that a 7 will appear are circled. Let us call this set B. Thus,

$$B = \{(1, 6), (2, 5), (3, 4), (4, 3), (5, 2), (6, 1) \}$$

Thus,

$$p(B) = \frac{6}{36} = \frac{1}{6}\}.$$

6.2 Relative Frequency: Large Numbers

The classical definition of probability includes the assumption that all possible outcomes are equally likely. Often, this is not the case. The statistical definition of probability is based on the notion of the relative frequency.

We define the statistical probability or empirical probability of an event as the relative frequency of occurrence of the event when the number of observations is very large. The probability is the limit of the relative frequency as the number of observations increases indefinitely.

EXAMPLE:

Suppose a coin was tossed 1,000 times and the result was 587 tails. The relative frequency of tails is $\dfrac{587}{1,000}$. Another 1,000 tosses lead to 511 tails. Then, the relative frequency of tails is

$$\frac{587+511}{1,000+1,000} = \frac{1,098}{2,000}.$$

Proceeding in this manner, we obtain a sequence of numbers, which gets closer and closer to the number defined as the probability of a tail in a single toss.

The empirical probability is based on the principle called the Law of Large Numbers.

6.2.1 The Law of Large Numbers

The sample mean tends to approach the population mean.

Here, by an event, we understand a subset of possible outcomes. It may contain none, one, some, or all of the possible outcomes.

Now we can define probability.

6.2.2 Definition of Probability

The probability of an event E is determined by associating 1 with the event occurring (success) and 0 with the event not occurring (failure). The experiment is performed a large number of times. The probability is defined as

$$\lim_{n\to\infty} \sum_{i=1}^{n} \frac{a_i}{n} = p = p(E)$$

where a_i is the outcome of the ith time the experiment is performed.

From the mathematical point of view this definition includes the concept of a limiting process, which may not exist. To avoid this trap, the axiomatic definition of probability was introduced.

6.3 Independent and Dependent Events: Conditional Probability

An event is a subset of all possible outcomes. Often, instead of saying event, we use the term set.

6.3.1   Union

The union of two sets, A and B, is the set of all elements that belong to A or to B. The union is denoted by $A \cup B$, read "A or B."

6.3.2   Intersection

The intersection of two sets, A and B, denoted by $A \cap B$, is the set containing all elements that belong to A and to B.

6.3.3   Difference

The difference of two sets, A and B, denoted by $A - B$, is the set of all elements of A that do not belong to B.

6.3.4   Complement

The complement of a set A, denoted by $\overline{A}$ or A^c or A', is the set of all elements (outcomes) in X that are not in A, where X represents the universal set.

6.3.5   Subset

A is a subset of B, denoted $A \subset B$, if every element of A is an element of B.

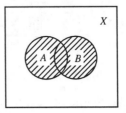

$A \subset B$ $A \cup B$ is shaded.

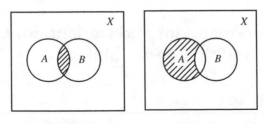

$A \cap B$ is shaded. $A - B$ is shaded.

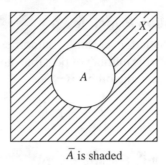

$\overline{A}$ is shaded

We will be using frequently the following identities:

1. $A \cup B = B \cup A$, $A \cap B = B \cap A$

2. $A \cup (B \cup C) = (A \cup B) \cup C$

    $A \cap (B \cap C) = (A \cap B) \cap C$

3. $A \cup (B \cap C) = (A \cup B) \cap (A \cup C)$

    $A \cap (B \cup C) = (A \cap B) \cup (A \cap C)$

4. $\overline{\overline{A}} = A$

5. $A - B = A \cap \overline{B}$

6.3.6 DeMorgan's Laws

$$\overline{A \cup B} = \overline{A} \cap \overline{B}$$
$$\overline{A \cap B} = \overline{A} \cup \overline{B}$$

In most cases, we assign an equal probability of $\dfrac{1}{n}$ to each of n possible outcomes. Sometimes, it is difficult to determine the value of n.

6.3.7 Multiplication Principle

If one experiment has n possible outcomes and another has m possible outcomes, then the number of possible outcomes of performing first one experiment, then the other, is

$$N = mn$$

6.3.8 Sampling with Replacement

Pick one of n balls from a bag and put it back. If the experiment is repeated m times, then the total number of possible outcomes is

$$N = n^m$$

6.3.9 Sampling without Replacement

Pick one of n balls from a bag and put it aside. Then, pick another ball from the bag. Repeat this m times (where $m \leq n$). The number of possible outcomes is

$$N = n(n-1)(n-2)(...)(n-m+1) = \frac{n!}{(n-m)!}$$

6.3.10 Conditional Probability

The probability that E occurs, given that F has occurred, is denoted by

$$P\,(E \mid F)$$

or $P(E$ given $F)$ and is called the conditional probability of E given that F has occurred.

We have

$$P(E|F) = \frac{P(E \cap F)}{P(F)}$$

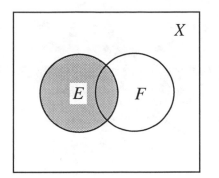 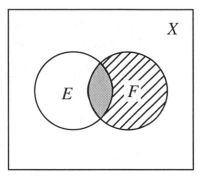

$P(E) =$

$$\frac{\text{shaded}}{\text{shaded} + \text{unshaded}}$$

$P(E|F) =$

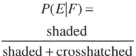

$$\frac{\text{shaded}}{\text{shaded} + \text{crosshatched}}$$

Often, we can find the probability of an intersection from

$$P(E \cap F) = P(E \mid F) \, P(F).$$

If

$$P(E \mid F) = P(E),$$

then we say that events E and F are independent; otherwise, they are dependent. If E and F are independent, then

$$P(E \cap F) = P(E) \times P(F).$$

For three events, E, F, and G, we have

$$P(E \cap F \cap G) = P(E) \; P(F \mid E) \, P(G \mid E \cap F).$$

If events E, F, and G are independent, then

$$P(E \cap F \cap G) = P(E) \times P(F) \times P(G).$$

One should not confuse independent events with mutually exclusive events. Two or more events are called mutually exclusive if the occurrence of any one of them excludes the occurrence of the others.

If E and F are independent, then

$$P(E \mid F) = P(E).$$

If E and F are mutually exclusive, then

$$P(E|F) = \frac{P(E \cap F)}{P(F)} = \frac{P(\emptyset)}{P(F)} = 0.$$

Let $E_1, E_2, \ldots, E_n$ be a partition of the set Ω of all outcomes, i.e.,

$$E_i \cap E_j = \emptyset \text{ for } i \neq j$$

and

$$\bigcup_{i=1}^{n} E_i = \Omega.$$

Then

$$p(E_1|E) = \frac{P(E_1)P(E|E_1)}{\displaystyle\sum_{i=1}^{n} P(E_n)P(E|E_n)}$$

This last equation is called Bayes' Theorem.

Problem Solving Examples:

Q There are two roads between towns A and B. There are three roads between towns B and C. How many different routes may one travel between towns A and C?

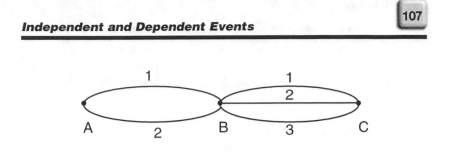

 If we take road 1 from town A to town B and then any road from B to C, there are three ways to travel from A to C. If we take road 2 from A to B and then any road from B to C, there are again three ways to travel from A to C. These two possibilities are the only ones available to us. Thus, there are $3 + 3 = 6$ ways to travel from A to C.

This problem illustrates the fundamental principle of counting. This principle states that if an event can be divided into k components and there are n_1 ways to carry out the first component, n_2 ways to carry out the second, n_i ways to carry out the ith, and n_k ways to carry out the kth, then there are $n_1 \times n_2 \times n_3 \times \ldots \times n_k$ ways for the original event to take place.

 Using only the digits 1, 2, 3, 4, 5, and 6, (a) how many different numbers containing three digits can be formed if any digit may be repeated? (b) How many different numbers are possible if no digit may be repeated?

 (a) There are six choices for each of the three digits of the chosen number. Then, $(6)(6)(6) = 6^3 = 216$ numbers.

(b) There are six choices for the first digit, five choices for the second digit, and four choices for the third digit. Then, $(6)(5)(4) = 120$ numbers.

 An ordinary die is tossed twice. What is the probability of getting a 2 on the first toss and an odd number on the second toss?

 These two events are independent.

Only one of the six sides is 2, so $P(2) = \dfrac{1}{6}$.

Since the only odd numbers are 1, 3, and 5, $P(\text{odd number}) = \dfrac{3}{6} = \dfrac{1}{2}$.

Thus, the final probability $= \left(\dfrac{1}{6}\right)\left(\dfrac{1}{2}\right) = \dfrac{1}{12}$.

 A committee is composed of six Democrats and five Republicans. Three of the Democrats are men, and three of the Republicans are men. If a man is chosen for chairperson, what is the probability that he is a Republican?

A Let E_1 be the event that a man is chosen, and E_2 the event that the man is a Republican.

We are looking for $P(E_2|E_1)$, From the definition of conditional probability $P(E_2|E_1) = \dfrac{P(E_1 \cap E_2)}{P(E_1)}$.

Of the 11 committee members, three are both male and Republican, hence,

$$P(E_1 \cap E_2) = \frac{\text{number of male Republicans}}{\text{number of committee members}} = \frac{3}{11}.$$

Of all the members, six are men (three Democrats and three Republicans); therefore,

$$P(E_1) = \frac{6}{11}.$$

Furthermore, $P(E_2|E_1) = \dfrac{P(E_1 \cap E_2)}{P(E_1)} = \dfrac{3/11}{6/11} = \dfrac{3}{6} = \dfrac{1}{2}.$

Q Twenty percent of the employees of a company are college graduates. Of these, 75% are in supervisory positions. Of those who did not attend college, 20% are in supervisory positions. What is the probability that a randomly selected supervisor is a college graduate?

Let the events be as follows:

E: The person selected is a supervisor.

E_1: The person is a college graduate.

E_2: The person is not a college graduate.

We are searching for $P(E_1|E)$.

By the definition of conditional probability

$$P(E_1|E) = \frac{P(E_1 \cap E)}{P(E)}.$$

But also by conditional probability $P(E_1 \cap E) = P(E|E_1)\,P(E_1)$. Since E is composed of mutually exclusive events, E_1 and E_2, $P(E) = P(E_1 \cap E) + P(E_2 \cap E)$. Furthermore, $P(E_2 \cap E) = P(E|E_2)$, by conditional probability. Inserting these expressions into $\dfrac{P(E_1 \cap E)}{P(E)}$, we obtain

$$P(E_1|E) = \frac{P(E_1)P(E|E_1)}{P(E_1)P(E|E_1) + P(E_2)P(E|E_2)}.$$

This formula is a special case of the well-known Bayes' Theorem. The general formula is

$$P(E_1|E) = \frac{P(E_1)P(E|E_1)}{\sum\limits_{1}^{n} P(En)P(E|E_n)}.$$

In our problem,

$P(E_1) = P(\text{college graduate}) = 20\% = .20$

$P(E_2) = P(\text{not a graduate}) = 1 - P(\text{graduate}) = 1 - .2 = .80$

$P(E|E_1) = P(\text{supervisor/graduate}) = 75\% = .75$

$P(E|E_2) = P(\text{supervisor/not a graduate}) = 20\% = .20$

Substituting,

$$P(E_1|E) = \frac{(.20)(.75)}{(.20)(.75) + (.80)(.20)} = \frac{.15}{.15 + .16} = \frac{15}{31}.$$

6.4   The Calculus of Probability

We denote

$\emptyset$ = the set of no outcomes and

Ω = the set of all possible outcomes (called certain event)

Here are the basic properties of probability:

$$P(\emptyset) = 0$$

$$P(\Omega) = 1$$

$$0 \leq P(E) \leq 1 \text{ for all events } E$$

$$P(E \cup F) = P(E) + P(F) - P(E \cap F)$$

If events E and F are mutually exclusive (i.e., $E \cap F = \emptyset$), then

$$P(E \cup F) = P(E) + P(F)$$

$$P(E \cup F \cup G) = P(E) + P(F) + P(G)$$

$$-P(E \cap F) - P(E \cap G) - P(F \cap G) + P(E \cap F \cap G)$$

In general, $P(E_1 \cup E_2 \cup \ldots \cup E_n) = P(E_1) + \ldots + P(E_n)$
$- P(\text{Intersection of twos}) + P(\text{Intersection of threes}) - \ldots \pm P(E_1 \cap \ldots \cap E_n)$.

$$P(\overline{E}) = 1 - P(E)$$

$$P(E \cap F) = P(E) + P(F) - P(E \cup F)$$

$$P(E \cap F) + P(E \cap \overline{F}) = P(E)$$

EXAMPLE:

The probability that A will be alive in ten years is 0.7, and the probability that B will be alive in ten years is 0.8. The probability that they both will be alive in ten years is

$$(0.7)\,(0.8) = 0.56.$$

EXAMPLE:

A die is tossed twice.

E = event "1, 2, or 3" on the first toss.

F = event "1, 2, 3, or 4" on the second toss.

We compute the probability of getting a 1, 2, or 3 on the first toss and a 1, 2, 3, or 4 on the second toss.

$$P(E) = \frac{3}{6} = \frac{1}{2}$$

$$P(F) = \frac{4}{6} = \frac{2}{3}$$

Events E and F are independent. Hence,

$$P(E \cap F) = \frac{1}{2} \times \frac{2}{3} = \frac{1}{3}.$$

EXAMPLE:

The probability of a 5 turning up at least once in two tosses of a die:

E = event "5" on the first toss.

F = event "5" on the second toss.

$E \cup F$ = event "5" on the first toss or "5" on the second toss or both.

We shall compute $P\,(E \cup F)$. E and F are not mutually exclusive, hence

$$P\,(E \cup F) = P\,(E) + P\,(F) - P\,(E \cap F).$$

Since E and F are independent we have

$$P\,(E \cap F) = P\,(E) \times P\,(F),$$

$$P(E \cup F) = P(E) + P(F) - P(E) \times P(F)$$

$$= \frac{1}{6} + \frac{1}{6} - \frac{1}{6} \times \frac{1}{6} = \frac{11}{36}.$$

EXAMPLE:

Two cards are drawn from a deck of 52 cards. Find the probability that they are both aces if the first card is:

1. replaced

2. not replaced

E = event "ace" on first draw, and

F = event "ace" on second draw.

1. If the first card is replaced, E and F are independent, then

$$P(E \cap F) = P(E)P(F) = \frac{4}{52} \times \frac{4}{52} = \frac{1}{169}.$$

2. The first card is drawn and not replaced. Both cards can be drawn in 52×51 ways. E can occur in four ways and F in three ways. Then

$$P(E \cap F) = \frac{4 \times 3}{52 \times 51} = \frac{1}{221}.$$

Another method:

$$P(E \cap F) = P(E)P(F|E) = \frac{4}{52} \times \frac{3}{51} = \frac{1}{221}.$$

EXAMPLE:

Find the probability of drawing either an ace or a club or both from a deck of cards.

E = event "drawing an ace"

F = event "drawing a club"

Note that events E and F are not mutually exclusive. Then

$$P(E \cup F) = P(E) + P(F) - P(E \cap F)$$
$$= \frac{4}{52} + \frac{13}{52} - \frac{1}{52} = \frac{4}{13}.$$

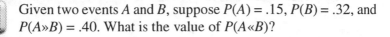

Problem Solving Examples:

Q Given two events A and B, suppose $P(A) = .15$, $P(B) = .32$, and $P(A»B) = .40$. What is the value of $P(A«B)$?

A $P(A \cup B) = P(A) + P(B) - P(A \cap B)$

$.40 = .15 + .32 - P(A \cap B)$

Let $P(A \cap B) = x$.

$$.40 = .15 + .32 - x$$

$$.40 = .47 - x \text{ or } x = .47 - .40$$

$$x = .07 = P(A \cap B)$$

Q Find the probability that on a single draw from a deck of playing cards we draw a spade or a face card or both. Define Events A and B as follows:

Event A: drawing a spade,

Event B: drawing a face card.

 We wish to find the probability of drawing a spade or a face card or both.

Let $A \cup B$ = the event of drawing a spade or face card or both.

$$P(A \cup B) = \frac{\text{number of ways a spade can occur}}{\text{total number of possible outcomes}}$$
$$+ \frac{\text{number of ways a face card can occur}}{\text{total number of possible outcomes}}$$

But we have counted some of the cards twice. Some cards are both spades and face cards, so we must subtract from the above expression the

$$-\frac{\text{number of ways a spade and face card can occur}}{\text{total number of possible outcomes}}$$

This can be rewritten as

$$P(A \cup B) = P(A) + P(B) - P(A \cap B).$$

$$P(A) = \frac{13}{52}, P(B) = \frac{12}{52}$$

$$P(A \cap B) = P(B)P(A|B).$$

$P(A|B)$ = Probability that a spade is drawn given that a face card is drawn.

$$= \frac{\text{number of spades that are face cards}}{\text{total number of face cards that could be drawn}}$$

$$= \frac{3}{12}$$

$$P(A \cap B) = \frac{12}{52} \times \frac{3}{12} - \frac{3}{52}$$

We could have found $P(A \cap B)$ directly by counting the number of spades that are face cards and then dividing by the total possibilities.

Thus, $P(A \cup B) = P(A) + P(B) - P(A \cap B)$

$$= \frac{13}{52} + \frac{12}{52} - \frac{3}{52} = \frac{22}{52} = \frac{11}{26}.$$

We could have found the answer more directly in the following way.

$$P(A \cup B) = \frac{\text{number of spades or face cards or both}}{\text{total number of cards}}$$

$$= \frac{22}{52} = \frac{11}{26}$$

6.5   Probability Distributions

6.5.1   Discrete Distributions

Variable X can assume a discrete set of values $x_1, x_2, \ldots, x_n$ with probabilities $p_1, p_2, \ldots, p_n$, respectively, where

$$p_1 + p_2 + \ldots + p_n = 1.$$

This defines a discrete probability distribution for X.

Probability function, or frequency function, is defined by

$$p : x_i \to p_i \quad i = 1, 2, \ldots, n$$
$$p(x_i) = p_i.$$

Variable X, which assumes certain values with given probabilities, is called a discrete random variable.

EXAMPLE:

A pair of dice is tossed. X denotes the sum of the points obtained, $X = 2, 3, \ldots, 12$. The probability distribution is given by

x	2	3	4	5	6	7	8	9	10	11	12
$p(x)$	$\frac{1}{36}$	$\frac{2}{36}$	$\frac{3}{36}$	$\frac{4}{36}$	$\frac{5}{36}$	$\frac{6}{36}$	$\frac{5}{36}$	$\frac{4}{36}$	$\frac{3}{36}$	$\frac{2}{36}$	$\frac{1}{36}$

$$\sum p(x) = 1$$

Replacing probabilities with relative frequencies, we obtain from the probability distribution a relative frequency distribution. Probability distributions are for populations, while relative frequency distributions are for samples drawn from this population.

The probability distribution, like a relative frequency distribution, can be represented graphically.

Cumulative probability distributions are obtained by cumulating probabilities. The function describing this distribution is called a distribution function.

EXAMPLE:

Find the probability of boys and girls in families with four children. Probabilities for boys and girls are equal.

$B =$ event "boy"

$G =$ event "girl"

$$P(B) = P(G) = \frac{1}{2}$$

1. Four boys

$$P(B \cap B \cap B \cap B) = P(B) \times P(B) \times P(B) \times P(B) = \frac{1}{16}$$

2. Three boys and one girl

$$P(B \cap B \cap B \cap G \cup B \cap B \cap G \cap B$$
$$\cup B \cap G \cap B \cap B \cup G \cap B \cap B \cap B)$$
$$= P(B) \times P(B) \times P(B) \times P(G) \times 4$$
$$= \frac{1}{2} \times \frac{1}{2} \times \frac{1}{2} \times \frac{1}{2} \times 4 = \frac{1}{4}$$

3. Three girls and one boy is the same as above

$$p = \frac{1}{4}$$

4. Two boys and two girls

$$P(B \cap B \cap G \cap G \cup B \cap G \cap B \cap G \cup B \cap G \cap G \cap B$$
$$\cup G \cap G \cap B \cap B \cup G \cap B \cap G \cap B \cup G \cap B \cap B \cap G)$$
$$= P(B) \times P(B) \times P(G) \times P(G) \times 6 = \frac{1}{16} = \frac{3}{8}$$

5. Four girls

$$p = \frac{1}{16}$$

Number of Girls x	4	3	2	1	0
Probability $p(X)$	$\frac{1}{16}$	$\frac{4}{16}$	$\frac{6}{16}$	$\frac{4}{16}$	$\frac{1}{16}$

Here, X is a random variable showing the number of girls in families with four children. The probability distribution is shown in the table.

This distribution can be represented graphically.

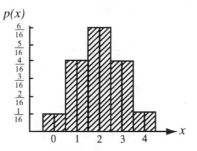

The sum of the areas of the rectangles is 1. Here, the discrete variable X is treated as a continuous variable. The figure is called a probability histogram.

6.5.2   Continuous Distributions

Suppose variable X can assume a continuous set of values. In such a case, the relative frequency polygon of a sample becomes (or rather tends to) a continuous curve.

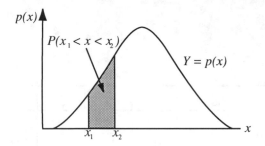

The total area under this curve is 1. The shaded area between the lines $x = x_1$ and $x = x_2$ is equal to the probability that x lies between x_1 and x_2.

Function $p(x)$ is a probability density function. Such a function defines a continuous probability distribution for X. The variable X is called a continuous random variable.

6.5.3 Mathematical Expectation

Let X be a discrete random variable which assumes the values $x_1, \ldots, x_n$ with respective probabilities $p_1, \ldots, p_n$ where $p_1 + p_2 + \ldots + p_n = 1$. The mathematical expectation of X denoted by $E(X)$ is defined as

$$E(X) = p_1 x_1 + \ldots + p_n x_n = \sum_{i=1}^{n} p_i x_i$$

Problem Solving Examples:

Q Let X be a continuous random variable. We wish to find probabilities concerning X. These probabilities are determined by a density function. Find a density function such that the probability that X falls in an interval (a, b) $(0 < a < b < 1)$ is proportional to the length of the interval (a, b). Check that this is a proper probability density function.

A The probabilities of a continuous random variable are computed from a continuous function called a density function in the following way. If $f(x)$ is graphed and is continuous,

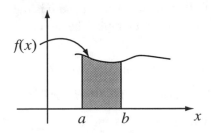

then $Pr(a \leq X \leq b)$ = the area under the curve $f(x)$ from a to b.

With this definition some conditions on $f(x)$ must be imposed. $f(x)$ must be positive and the total area between $f(x)$ and the x-axis must be equal to 1.

We also see that if probability is defined in terms of area under a curve, the probability that a continuous random variable is equal to a particular value, $Pr(X = a)$ is the area under $f(x)$ at the point a. The area of a line is 0, thus, $Pr(X = a) = 0$. Therefore,

$$Pr(a < X < b) = Pr(a \leq X \leq b).$$

To find a density function for $0 < X < 1$, such that $Pr(a < X < b)$ is proportional to the length of (a, b), we look for a function $f(x)$ that is positive and the area under $f(x)$ between 0 and 1 is equal to 1. It is reasonable to expect that the larger the interval the larger the probability that x is in the interval.

A density function that satisfies these criteria is

$$f(x) = \begin{cases} 1 & 0 \leq x \leq 1 \\ 0 & \text{otherwise.} \end{cases}$$

A graph of this density function is

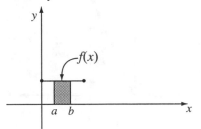

The probability that X is between a and b is the area of the shaded region. This is the area of a rectangle. The area of a rectangle is base × height.

Thus, $Pr(a \leq X \leq b) = (b - a) \times 1 = b - a$.

Similarly, $Pr(X \leq k) = (k - 0) \times 1 = k$ for $0 < k < 1$.

Often the density function is more complicated, and integration must be used to calculate the area under the density function.

To check that this is a proper probability density function, we must check that the total area under $f(x)$ is 1. The total area under this density function is $(1 - 0) \times 1 = 1$.

 Given that the random variable X has density function

$$f(x) = \begin{cases} 2x & 0 < x < 1 \\ 0 & \text{otherwise} \end{cases}$$

Find $Pr = \left(\dfrac{1}{2} < x < \dfrac{3}{4} \right)$ and $Pr\left(-\dfrac{1}{2} < x < \dfrac{1}{2} \right)$

A Since $f(x) = 2x$ is the density function of a continuous random variable, $Pr = \left(\dfrac{1}{2} < x < \dfrac{3}{4} \right)$ = area under $f(x)$ from $\dfrac{1}{2}$ to $\dfrac{3}{4}$.

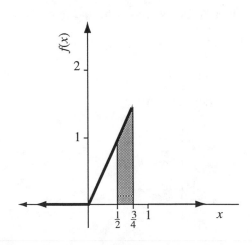

$f(x)$ is indicated by the heavy line.

The area under $f(x)$ is the area of the triangle with vertices at $(0, 0)$, $(1, 0)$, and $(1, 2)$.

The area of this triangle is $A = \dfrac{1}{2} bh$, where b is the base of the triangle and h is the altitude.

Thus, $A = \dfrac{1}{2}(1) \times 2 = \dfrac{2}{2} = 1$

proving that $f(x)$ is a proper probability density function.

To find the probability that $\dfrac{1}{2} < x < \dfrac{3}{4}$, we find the area of the shaded region in the diagram. This shaded region is the difference in areas of the right triangle with vertices $(0, 0)$, $\left(\dfrac{1}{2}, 0\right)$, and $\left(\dfrac{1}{2}, f\left(\dfrac{1}{2}\right)\right)$

and the area of the triangle with vertices $(0, 0)$, $\left(\dfrac{3}{4}, 0\right)$, and $\left(\dfrac{3}{4}, f\left(\dfrac{3}{4}\right)\right)$.

This difference is

$$\Pr\left(\frac{1}{2} < x < \frac{3}{4}\right) = \frac{1}{2}\left(\frac{3}{4}\right) f\left(\frac{3}{4}\right) - \frac{1}{2}\left(\frac{1}{2}\right) f\left(\frac{1}{2}\right)$$

$$= \frac{1}{2}\left[\frac{3}{4} \times \frac{6}{4} - \frac{1}{2} \times 1\right]$$

$$= \frac{1}{2}\left[\frac{9}{8} - \frac{1}{2}\right] = \frac{1}{2} \times \frac{5}{8} = \frac{5}{16}.$$

The probability that $-\dfrac{1}{2} < x < \dfrac{1}{2}$ is

$$Pr\left(-\frac{1}{2} < x < \frac{1}{2}\right) = \text{Area under } f(x) \text{ from } -\frac{1}{2} \text{ to } \frac{1}{2}.$$

Because $f(x) = 0$ from $-\dfrac{1}{2}$ to 0, the area under $f(x)$ from $-\dfrac{1}{2}$ to 0 is 0. Thus,

$$\Pr\left(-\frac{1}{2} < x < \frac{1}{2}\right) = \Pr\left(0 < x < \frac{1}{2}\right) = \text{area under } f(x) \text{ from 0 to } \frac{1}{2}$$

$$= \frac{1}{2}\left(\frac{1}{2}\right)f\left(\frac{1}{2}\right)$$

$$= \frac{1}{2}\left(\frac{1}{2}\right) \times 1 = \frac{1}{4}$$

 Let X be the random variable defined as the number of dots observed on the upturned face of a fair die after a single toss. Find the expected value of X.

 X can take on the values 1, 2, 3, 4, 5, or 6. Since the die is fair, we assume that each value is observed with equal probability. Thus,

$$\Pr(X = 1) = \Pr(X = 2) = \ldots = \Pr(X = 6)$$

$$= \frac{1}{6}$$

The expected value of X is

$E(X) = \sum x \Pr(X = x).$ Hence,

$$E(X) = 1 \times \frac{1}{6} + 2 \times \frac{1}{6} + 3 \times \frac{1}{6} + 4 \times \frac{1}{6} + 5 \times \frac{1}{6} + 6 \times \frac{1}{6}$$

$$= \frac{1}{6}(1 + 2 + 3 + 4 + 5 + 6)$$

$$= \frac{21}{6} = 3\frac{1}{2}.$$

Suppose the earnings of a laborer, denoted by X, are given by the following probability function.

X	0	8	12	16
$Pr(X = x)$	0.3	0.2	0.3	0.2

Find the laborer's expected earnings.

 The laborer's expected earnings are denoted by $E(X)$, the expected value of the random variable X.

The expected value of X is defined to be

$E(X) = (0)\, Pr(X = 0) + (8)\, Pr(X = 8)$

$\qquad + (12)\, Pr(X = 12) + (16)\, Pr(X = 16)$

$\qquad = (0)(.3) + (8)(.2) + (12)(.3) + (16)(.2)$

$\qquad = 0 + 1.6 + 3.6 + 3.2 = 8.4$

Thus, the expected earnings are 8.4.

6.6   Combinational Analysis

If an event can happen in any one of n ways, and if, when this has occurred, another event can happen in any one of m ways, then the number of ways in which both events can happen in the specified order is

$$n \times m.$$

EXAMPLE:

A bank offers three types of checking accounts and four types of savings accounts. Any customer who wants to have both accounts has $3 \times 4 = 12$ possibilities.

6.6.1   Definition of Factorial

Factorial n, denoted by $n!$, is defined as

$$n! = 1 \times 2 \times \dots \times (n - 1) \times n \qquad\qquad 0! = 1$$

EXAMPLE:

$6! = 1 \times 2 \times 3 \times 4 \times 5 \times 6 = 720$

$1! = 1, \quad 2! = 2$

EXAMPLE:

In how many ways can six children be arranged in a row?

Each of six children can fill the first position. Then, each of five remaining children can fill the second place. Thus, there are 6×5 ways of filling the first two places. There are four ways of filling the third place, three ways of filling the fourth place, two ways of filling the fifth place, and one way of filling the sixth place. Hence,

Number of arrangements $= 6 \times 5 \times 4 \times 3 \times 2 \times 1 = 6! = 720$

We have a general rule: number of arrangements of n different objects in a row is $n!$.

6.6.2   Permutations

The number of ways k objects can be picked from among n objects with regard to order is denoted by $_nP_k$ or $P(n, k)$ or $P_{n,\,k}$ and is equal to

$$_nP_k = n(n-1)(n-2)\ldots(n-k+1) = \frac{n!}{(n-k)!}$$

EXAMPLE:

Three letters are given x, y, and z. There are $_3P_2$ ways of choosing two letters at a time:

$$_3P_2 = \frac{3!}{(3-2)!} = 2 \times 3 = 6$$

Indeed, there are six possibilities:

$$xy,\ xz,\ yx,\ yz,\ zx,\ zy$$

EXAMPLE:

From the set of numbers $1, 2, 3, 4, 5, 6, 7, 8,$ and 9, one chooses three at a time. There are

$$_9P_3 = \frac{9!}{6!} = 7 \times 8 \times 9 = 504$$

three-digit numbers.

Observe that the number of permutations of n objects, taken n at a time, is

$$_nP_n = \frac{n!}{(n-n)!} = n!$$

$_nP_k$ is called the number of permutations of n different objects, taken k at a time.

EXAMPLE:

In how many ways can ten people be arranged in a circle?

Person number one can be placed anywhere. The remaining nine people can be arranged in

$$9! = 362,880$$

ways.

The number of permutations of n objects consisting of groups in which n_1 are alike, n_2 are alike, ... is

$$\frac{n!}{n_1! - n_2! \dots}$$

where $n_1 + n_2 + \dots = n$.

EXAMPLE:

The number of permutations of letters in the word crisscross is

$$\frac{10!}{4!2!2!1!1!} = 37,800.$$

6.6.3 Combinations

The number of ways one can choose k objects out of the n objects, disregarding the order, is denoted by

$$_nC_k, C(n,k) \text{ or } \binom{n}{k}$$

and is equal to

$$_nC_k = \frac{n(n+1)\ldots(n-k+1)}{k!} = \frac{n!}{k!(n-k)!} = \frac{_nP_k}{k!}$$

$_nC_k$ is the number of combinations of n objects taken k at a time.

EXAMPLE:

The number of combinations of the letters x, y, and z, taken two at a time, is

$$_3C_2 = \frac{3!}{2!(3-2)!} = 3$$

The combinations are xy, xz, and yz. Combinations xy and yx are the same, but permutations xy and yx are not the same.

EXAMPLE:

A committee consists of seven people. There are 13 candidates. In how many ways can the committee be chosen?

$$_{13}C_7 = \frac{13!}{7!6!} = \frac{8 \times 9 \times 10 \times 11 \times 12 \times 13}{2 \times 3 \times 4 \times 5 \times 6} = 1,716$$

EXAMPLE:

How many possible outcomes has a lotto game where a player chooses seven numbers out of 49 numbers?

$$_{49}C_7 = \frac{49!}{7!42!} = \frac{43 \times 44 \times 45 \times 46 \times 47 \times 48 \times 49}{2 \times 3 \times 4 \times 5 \times 6 \times 7}$$
$$= 85,900,584$$

Note: $_nC_k = {_n}C_{n-k}$.

For example, $_{49}C_7 = \dfrac{49!}{7!42!} = \dfrac{49!}{42!7!} = {_{49}}C_{42}$

The number of combinations of n objects taken 1, 2, ..., n at a time is

$$_nC_1 + {_n}C_2 + ... + {_n}C_n = 2^n - 1$$

It is difficult to evaluate $n!$ for large numbers. In such cases, an approximate formula (called Stirling's Formula) is used:

$$n! \approx \sqrt{2\pi n}\, n^n e^{-n}$$

where e is the natural base of logarithms, $e = 2.718281828...$

EXAMPLE:

Determine the probability of four 4's in six tosses of a die. The result of each toss is the event $\overline{4}$ or non 4 (4). Thus,

$$4,4,4,\overline{4},4,\overline{4}$$

or

$$\overline{4},4,4,\overline{4},4,4$$

are successes.

The probability of an event $4,4,4,\overline{4},4,\overline{4}$ is

$$P(4,4,4,\overline{4},4,\overline{4}) = \frac{1}{6} \times \frac{1}{6} \times \frac{1}{6} \times \frac{5}{6} \times \frac{1}{6} \times \frac{5}{6}$$

$$= \left(\frac{1}{6}\right)^4 \times \left(\frac{5}{6}\right)^2$$

All events in which four 4's and two non 4's occur have the same probability. The number of such events is

$$_6C_4 = \frac{6!}{4!2!} = 15$$

and all these events are mutually exclusive. Hence,

$$P(\text{four 4's in 6 tosses}) = 15 \times \left(\frac{1}{6}\right)^4 \times \left(\frac{5}{6}\right)^2 = 0.008$$

EXAMPLE:

Thirty percent of the cars produced by a factory have some defect. A sample of 100 cars is selected at random. What is the probability that

1. exactly 10 cars will be defective?

2. 95 or more will be defective?

In general, if $p = p(E)$ and $q = p(\overline{E})$, then the probability of getting exactly m E's in n trials is

$$_cC_m p^m q^{n-m}$$

1. $p(10 \text{ defective cars}) = {}_{100}C_{10} \left(\frac{3}{10}\right)^{10} \left(\frac{7}{10}\right)^{90} \approx .00000117$

2. $p(95 \text{ or more defective}) = p(95 \text{ defective}) +$

$p(96 \text{ defective}) + p(97 \text{ defective}) + p(98 \text{ defective}) +$

$p(99 \text{ defective}) + p(100 \text{ defective}) =$

$$={}_{100}C_{95}\left(\frac{3}{10}\right)^{95}\left(\frac{7}{10}\right)^{5} + {}_{100}C_{96}\left(\frac{3}{10}\right)^{96}\left(\frac{7}{10}\right)^{4}$$

$$+\ldots+ {}_{100}C_{100}\left(\frac{3}{10}\right)^{100}\left(\frac{7}{10}\right)^{0} \approx 2.43\times10^{-43}$$

Problem Solving Examples:

 Using all 26 letters of the alphabet, in how many ways can four letters be chosen?

 The number of permutations of 26 items taken four at a time = (26)(25)(24)(23), which is also ${}_{26}P_4 = \dfrac{26!}{(26-4)!} = \dfrac{26!}{22!}$

Note: 26! = (26)(25)(24)(…)(1) and 22! = (22)(21)(20)(…)(1)

Solving $\dfrac{26!}{22!} = 358{,}800$.

 Determine the number of permutations of the letters in the word BANANA.

 In solving this problem we use the fact that the number of permutations P of n things taken all at a time [$P(n, n)$], of which n_1 are alike, n_2 others are alike, n_3 others are alike, etc. is

$$P = \frac{n!}{n_1!n_2!n_3!\ldots}, \text{ with } n_1 + n_2 + n_3 + \ldots = n$$

In the given problem there are six letters ($n = 6$), of which two are alike (there are two N's so that $n_1 = 2$), three others are alike (there are three A's, so that $n_2 = 3$), and one is left (there is one B, so $n_3 = 1$). Notice that $n_1 + n_2 + n_3 = 2 + 3 + 1 = 6 = n$; thus,

$$P = \frac{6!}{2!3!1!} = \frac{6 \times 5 \times \overset{2}{4} \times 3}{2 \times 1 \times 3! \times 1} = 60.$$

Thus, there are 60 permutations of the letters in the word BANANA.

 How many baseball teams of nine members can be chosen from among 12 players, without regard to the position played by each member?

 Since there is no regard to position, this is a combinations problem (if order or arrangement had been important, it would have been a permutations problem). The general formula for the number of combinations of n things taken r at a time is

$$C(n,r) = \frac{n!}{r!(n-r)!}.$$

We have to find the number of combinations of 12 things taken nine at a time. Hence, we have

$$C(12,9) = \frac{12!}{9!(12-9)!} = \frac{12!}{9!3!} = \frac{12 \times 11 \times 10 \times 9!}{3 \times 2 \times 1 \times 9!} = 220.$$

Therefore, there are 220 possible teams.

 What is the probability of getting exactly four 6's when a die is rolled seven times?

 Let X = the number of 6's observed when a die is rolled seven times. If we assume that each roll is independent of each other roll and that the probability of rolling a 6 on one roll is $= \frac{1}{6}$, the X is

binomially distributed with parameters $n = 7$ and $p = \frac{1}{6}$.

Thus, $Pr(X = 4) = Pr$(exactly four 6's on seven rolls)

$$=_7C_4 = \left(\frac{1}{6}\right)^4 \left(\frac{5}{6}\right)^{7-4}$$

$$= \frac{7 \times 6 \times 5 \times 4 \times 3 \times 2 \times 1}{4 \times 3 \times 2 \times 3 \times 2 \times 1} \left(\frac{1}{6}\right)^4 \left(\frac{5}{6}\right)^3$$

$$= 35 \left(\frac{1}{6}\right)\left(\frac{5}{6}\right)$$

$$Pr(X = 4) = 35 \left(\frac{1}{1,296}\right)\left(\frac{125}{216}\right) = \frac{4,375}{279,936} = .0156.$$

 If the probability of your hitting a target on a single shot is .8, what is the probability that in four shots you will hit the target at least twice?

Each shot at the target is an independent trial with constant probability, $p = .8$, of a success. The only other possibility is failure. This type of situation calls for the binomial distribution,

$$P(X = k \text{ successes}) = \left(\frac{n}{k}\right) p^k (1-p)^{n-k}.$$

Since the events of 2, 3, or 4 successes are mutually disjointed, we use the addition rule for probabilities and

$$P(2 \text{ or } 3 \text{ or } 4) = P(2) + P(3) + P(4)$$

$$P(X = 2) = \frac{4!}{2!(4-2)!}(.8)^2(.2)^{4-2} = \frac{4!}{2!2!}(.8)^2(.2)^2$$

$$= \frac{4 \times 3 \times 2 \times 1}{2 \times 1 \times 2 \times 1}(.64)(0.04)$$

$$= 6(.0256) = 0.1536.$$

$$P(X=3) = \frac{4!}{3!(4-3)!}(.8)^3(.2)^{4-3} = \frac{4!}{3!1!}(.8)^3(.2)$$

$$= 4(0.512)(0.2) = 0.4096.$$

$$P(X=4) = \frac{4!}{4!(4-4)!}(.8)^4(.2)^{4-4} = \frac{4!}{4!0!}(.8)^4(.2)^0$$

$$= (0.8)^4 \text{ since } \frac{4!}{4!} = 1 \text{ and } 0! = 1 \text{ (by definition) and any}$$

number raised to the zero power = 1.

$$= 0.4096.$$

Thus, $P(2 \text{ or } 3 \text{ or } 4) = 0.1536 + 0.4096 + 0.4096 = 0.9728$

6.7 Set Theory and Probability

First, we define a sample space X, which consists of all possible outcomes of an experiment. With each point of X, we can associate a non-negative number called a probability. If X contains only a finite number of points, then the sum of all probabilities is equal to one.

An event is defined as a set of points in X.

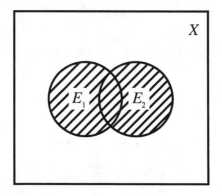

Next, we define $E_1 \cup E_2, E_1 \cap E_2, E_1 - E_2$, and mutually exclusive events (i.e, events E_1 and E_2 such that $E_1 \cap E_2 = \emptyset$). The probability associated with the null set is zero:

$$p(\emptyset) = 0$$

EXAMPLE:

A die is tossed twice. Using a sample space, we find the probability that the sum of two tosses is either three or seven. The sample space consists of all points $(1, 1)$, $(1, 2)$, ..., $(6, 6)$.

$(1, 1)$ $(2, 1)$ $(3, 1)$ $(4, 1)$ $(5, 1)$ $(6, 1)$

$(1, 2)$ $(2, 2)$ $(3, 2)$ $(4, 2)$ $(5, 2)$ $(6, 2)$

$(1, 3)$ $(2, 3)$ $(3, 3)$ $(4, 3)$ $(5, 3)$ $(6, 3)$

$(1, 4)$ $(2, 4)$ $(3, 4)$ $(4, 4)$ $(5, 4)$ $(6, 4)$

$(1, 5)$ $(2, 5)$ $(3, 5)$ $(4, 5)$ $(5, 5)$ $(6, 5)$

$(1, 6)$ $(2, 6)$ $(3, 6)$ $(4, 6)$ $(5, 6)$ $(6, 6)$

The sample space consists of 36 points. To each point, we assign probability $\dfrac{1}{36}$:

$$36 \times \frac{1}{36} = 1$$

A = event "sum equals three"

B = event "sum equals seven"

$$p(A) = \frac{2}{36}, p(B) = \frac{6}{36}$$

A and B have no points in common, that is, they are mutually exclusive. Thus,

$$p(A \cup B) = p(A) + p(B) = \frac{8}{36} = \frac{2}{9}.$$

Problem Solving Examples:

 If a card is drawn from a deck of playing cards, what is the probability that it will be a jack or a ten?

 The probability that an event A or B occurs, but not both at the same time, is $P(A \cup B) = P(A) + P(B)$. Here the symbol "$\cup$" stands for "or."

In this particular example, we select only one card at a time. Thus, we either choose a jack "or" a ten. $P(\text{a jack or a ten}) = P(\text{a jack}) + P(\text{a ten})$.

$$P(\text{a jack}) = \frac{\text{number of ways to select a jack}}{\text{number of ways to choose a card}} = \frac{4}{52} = \frac{1}{13}.$$

$$P(\text{a ten}) = \frac{\text{number of ways to select a ten}}{\text{number of ways to choose a card}} = \frac{4}{52} = \frac{1}{13}.$$

$$P(\text{a jack or a ten}) = P(\text{a jack}) + P(\text{a ten}) = \frac{1}{13} + \frac{1}{13} = \frac{2}{13}.$$

 Determine the probability of getting 6 or 7 in a toss of two dice.

 Let A = the event that a 6 is obtained in a toss of two dice

and B = the event that a 7 is obtained in a toss of two dice.

Then, the probability of getting 6 or 7 in a toss of two dice is

$$P(A \text{ or } B) = P(A \cup B).$$

The union symbol "$\cup$" means that A and/or B can occur. Now $P(A \cup B) = P(A) + P(B)$ if A and B are mutually exclusive. Two or more events are said to be mutually exclusive if the occurrence of any one of them excludes the occurrence of the others. In this case, we cannot obtain a six and a seven in a single toss of two dice. Thus, A and B are mutually exclusive.

To calculate $P(A)$ and $P(B)$, use the following table. Note: There are 36 different tosses of two dice.

$A = 6$ is obtained in a toss of two dice.

$= \{(1, 5), (2, 4), (3, 3), (4, 2), (5, 1)\}$

$B = 7$ is obtained in a toss of two dice.

$= \{(1, 6), (2, 5), (3, 4), (4, 3), (5, 2), (6, 1)\}.$

$$P(A) = \frac{\text{number of ways to obtain a 6 in a toss of two dice}}{\text{number of ways to toss two dice}}$$

$$= \frac{5}{36}$$

$$P(B) = \frac{\text{number of ways to obtain a 7 in a toss of two dice}}{\text{number of ways to toss two dice}}$$

$$= \frac{6}{36} = \frac{1}{6}$$

Quiz: Measures of Variability – Probability Theory

1. Which of the following four sets of data have the smallest and largest standard deviations?

I	II	III	IV
1	1	1	1
2	5	1	3
3	5	1	3
4	5	1	3
5	5	5	5
6	5	9	7
7	5	9	7
8	5	9	7
9	9	9	9

Smallest/largest:

(A) II/I

(D) II/IV

(B) III/I

(E) II/III

(C) IV/I

2. Polly takes three standardized tests. She scores 600 on all three. Using standard scores, or z-scores, rank her performance on the three tests from best to worst if the means and standard deviations for the tests are as follows:

	Mean	**Standard Deviation**
Test I	500	80
Test II	470	120
Test III	560	30

Her rankings are

(A) I, II, III.

(D) I, III, II.

(B) III, II, I.

(E) III, I, II.

(C) II, I, III.

3. The graph shows the distribution of scores on a standardized mental test. A school counselor deals exclusively with children with exceptional abilities and those who are mentally challenged. He knows that a child who walks into his office has already obtained a score that falls in one of the shaded regions in the graph. If Albert appears in the counselor's office, he must be among

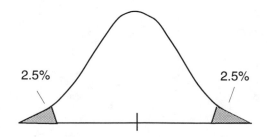

    (A)   the top 2.5% of the population.

    (B)   the bottom 2.5% of the population.

    (C)   the extreme 5% of the population.

    (D)   the middle 95% of the population.

    (E)   the top 97.5% of the population.

4.   The coefficient of variations for a sample is .40. If the standard deviation is 5, what is the value of the mean?

    (A) 12.5   (D) 4.5

    (B) 10   (E) 2

    (C) 7.5

5.   How many games would it take a baseball coach to try every possible batting order with nine players?

    (A) 9   (D) 362,880

    (B) 45   (E) 3.8742×10^8

    (C) 81

6.   In how many different ways can the letters *a, b, c,* and *d* be arranged if they are selected three at a time?

    (A) 8   (D) 4

    (B) 12   (E) 48

    (C) 24

7. What is the probability that in a single throw of two dice the sum of 10 will appear?

    (A) $\dfrac{10}{36}$ (D) $\dfrac{2}{10}$

    (B) $\dfrac{1}{6}$ (E) $\dfrac{11}{12}$

    (C) $\dfrac{1}{12}$

8. A bag contains four white balls, six black balls, three red balls, and eight green balls. If one ball is drawn from the bag, find the probability that it will be either white or green.

    (A) $\dfrac{1}{3}$ (D) $\dfrac{4}{13}$

    (B) $\dfrac{2}{3}$ (E) $\dfrac{8}{21}$

    (C) $\dfrac{4}{7}$

9. An urn contains 30 blue balls, 40 green balls, and 15 red balls. What is the probability of choosing a red ball first followed by a blue ball with no replacement?

    (A) 0.3571   (D) 0.0620

    (B) 0.1765   (E) 0.0630

    (C) 0.3529

10. Six dice are thrown. What is the probability of getting six 1's?

    (A) 0.0000214   (D) 0.1667

    (B) 0.0278   (E) 0.1

    (C) 0.00001

ANSWER KEY

1.	(E)	6.	(C)
2.	(E)	7.	(C)
3.	(C)	8.	(C)
4.	(A)	9.	(E)
5.	(D)	10.	(A)

CHAPTER 7

Distributions

7.1 The Binomial Distribution

Let E be an event and p be the probability that E will happen in any single trial. Number p is called the probability of a success. Then $q = 1 - p$ is the probability of a failure.

The probability that the event E will happen exactly x times in n trials is given by

$$p(x) =_n C_x p^x (1-p)^{n-x} = \frac{n!}{x!(n-x)!} p^x (1-p)^{n-x} \quad (1)$$

where $p(x)$ is the probability of x successes and $n - x$ failures.

The range of x is $0, 1, 2, \ldots, n$.

Distribution (1) is called the Bernoulli distribution or the binomial distribution. For $x = 0, 1, \ldots, n$, we obtain in (1) $_n C_0, _n C_1, _n C_2, \ldots, _n C_n$ which are the binomial coefficients of the binomial expansion

$$(p+q)^n =_n C_o p^n +_n C_1 p^{n-1}q +_n C_2 p^{n-2}q^2 + \ldots +_n C_n q^n.$$

EXAMPLE:

Toss a die five times. An event E is that a 6 appears. We find the binomial distribution.

$$p = \frac{1}{6}, q = 1 - p = \frac{5}{6}, n = 5$$

$$p(0) = {}_5 C_0 \left(\frac{1}{6}\right)^0 \left(\frac{5}{6}\right)^5 = \left(\frac{5}{6}\right)^5 = 0.40187$$

$$p(1) = {}_5 C_1 \left(\frac{1}{6}\right) \left(\frac{5}{6}\right)^4 = 0.40187$$

$$p(2) = {}_5 C_2 \left(\frac{1}{6}\right)^2 \left(\frac{5}{6}\right)^3 = 0.16075$$

$$p(3) = {}_5 C_3 \left(\frac{1}{6}\right)^3 \left(\frac{5}{6}\right)^2 = 0.03215$$

$$p(4) = {}_5 C_4 \left(\frac{1}{6}\right)^4 \left(\frac{5}{6}\right)^1 = 0.00321$$

$$p(5) = {}_5 C_5 \left(\frac{1}{6}\right)^5 \left(\frac{5}{6}\right)^0 = 0.00013$$

EXAMPLE:

Evaluate the expectation of x, i.e.,

$$\sum_{x=0}^{n} x\, p(x), \quad \text{for } p(x) = {}_n C_x p^x q^{n-x}$$

$$\sum_{x=0}^{n} x\, p(x) = \sum_{x=1}^{n} x \frac{n!}{x!(n-1)!} p^x q^{n-x} \qquad (2)$$

$$= np \sum_{x=1}^{n} \frac{(n-1)!}{(x-1)!(n-x)!} p^{x-1} q^{n-x}$$

$$= np(p+q)^{n-1} = np$$

since $p + q = 1$.

EXAMPLE:

Evaluate the expectation of x^2, i.e., $\sum\limits_{x=0}^{n} x^2 p(x)$ where $p(x)$ is a binomial distribution.

$$\sum_{x=0}^{n} x^2 p(x) = \sum_{x=1}^{n} x^2 \frac{n!}{x!(n-x)!} p^x q^{n-x}$$

$$= \sum_{x=1}^{n} [x(x-1)+x] \frac{n!}{x!(n-x)!} p^x q^{n-x} \qquad (3)$$

$$= \sum_{x=2}^{n} x(x-1) \frac{n!}{x!(n-x)!} p^x q^{n-x}$$

$$+ \sum_{x=1}^{n} x \frac{n!}{x!(n-x)!} p^x q^{n-x}$$

$$= n(n-1) p^2 \sum_{x=2}^{n} \frac{(n-2)!}{(x-2)!(n-x)!} p^{x-2} q^{n-x} + np$$

$$= n(n-1) p^2 (p+q)^{n-2} + np = n(n-1) p^2 + np$$

Next, we compute mean μ and variance σ^2 of a binomially distributed variable.

$$\mu = \sum_{x=0}^{n} x p(x) = np$$

$$\sigma^2 = \sum_{x=0}^{n} (x-\mu)^2 p(x) = \sum_{x=0}^{n} (x^2 - 2\mu x + \mu^2) p(x)$$

$$= \sum_{x=0}^{n} x^2 p(x) - 2\mu \sum_{x=0}^{n} x p(x) + \mu^2 \sum_{x=0}^{n} p(x)$$

$$= n(n-1) p^2 + np - 2\mu^2 + \mu^2$$

$$= np(1-p) = npq$$

EXAMPLE:

A coin is tossed 10,000 times. The mean number of heads is

$$\mu = np = 10,000 \frac{1}{2} = 5,000$$

In 10,000 tosses 5,000 heads are expected. The standard deviation is

$$\sigma = \sqrt{npq} = 100 \times \frac{1}{2} = 50$$

Properties of the Binomial Distribution

Mean	$\mu = np$
Standard deviation	$\sigma = \sqrt{npq}$
Variance	$\sigma^2 = npq$
Moment coefficient of skewness	$a_3 = \dfrac{q-p}{\sqrt{npq}}$
Moment coefficient of kurtosis	$a_4 = 3 + \dfrac{1-6pq}{npq}$

Observe that the binomial distribution is a sum of n distributions defined by

$$f(1) = p \qquad x \, \varepsilon \, \{0, 1\}$$

Where E means "belongs to." In this case, x must be 0 or 1, so x belongs to the set containing 0, 1.

$$f(0) = q = 1 - p$$

$$\text{Mean} \qquad \mu = p$$

$$\text{Variance} \qquad \sigma^2 = p(1-p) = pq$$

Some other distributions are also used, such as: multinomial, hypergeometric, and geometric.

Problem Solving Examples:

 For a given coin, the probability of getting tails when it is tossed is .6. If this coin is tossed 900 times, what is the expected (mean) number of tails?

 m = mean = np, where n = number of tosses and p = probability of tails.

Mean number of tails = (900) (.6)

$$\mu = 540.$$

 In the previous question, what is the standard deviation (s) of the distribution?

 $\sigma = \sqrt{npq}$, where n = number of tosses, p = probability of tails, q = probability of heads, and σ = standard deviation.

$$\sigma = \sqrt{(900)(.6)(.4)}$$

Solving,

$$= \sqrt{(216)} \approx 14.7.$$

7.2   The Multinomial Distribution

The probability of events $E_1, E_2, \ldots, E_k$ are $p_1, p_2, \ldots, p_k$, respectively. The probability that $E_1, \ldots, E_k$ will occur $x_1, \ldots, x_k$ times, respectively, is

$$\frac{n!}{x_1 ! x_2 ! \ldots x_k !} \, p_1^{x_1} p_2^{x_2} \ldots p_k^{x_k} \qquad (4)$$

where $x_1 + x_2 + \ldots + x_k = n$. Distribution (4) is called the multinomial distribution. The expected number of times each event $E_1, \ldots, E_k$ will occur in n trials are $np_1, \ldots, np_k$, respectively.

$$np_1 + \ldots + np_k = n(p_1 + \ldots + p_k) = n.$$

Observe that (4) is the general term in the multinomial expansion $(p_1 + \ldots + p_k)^n$.

EXAMPLE:

A die is tossed ten times. The probability of getting 1, 2, 3, 4, and 5 exactly once and 6 five times is

$$\frac{10!}{5!}\left(\frac{1}{6}\right)\left(\frac{1}{6}\right)\left(\frac{1}{6}\right)\left(\frac{1}{6}\right)\left(\frac{1}{6}\right)\left(\frac{1}{6}\right)^5 = \frac{7 \cdot 8 \cdot 9 \cdot 10}{6^9}$$

$$= 0.0005.$$

Problem Solving Example:

Q A die is tossed 12 times. Let xi denote the number of tosses in which i dots come up for i = 1, 2, 3, 4, 5, and 6. What is the probability that we obtain two of each value (i.e., two 1's, two 2's, two 3's, etc.)?

A We have a series of independent successive trials with six possible outcomes each with constant probability $\frac{1}{6}$. The multinomial distribution,

$$p(x_1 = f_1, f_2, = f_2, \ldots x_k = f_k)$$

$$= \frac{n!}{f_1! f_2! , \ldots f_k!} p_1^{f_1} p_2^{f_2} \ldots p_k^{f_k},$$

is called for. Hence,

$$p(x_1 = 2, x_2 = 2, x_3 = 2, x_4 = 2, x_5 = 2, x_6 = 2)$$

$$= \frac{12!}{2!2!2!2!2!2!}\left(\frac{1}{6}\right)\left(\frac{1}{6}\right) \ldots \left(\frac{1}{6}\right)$$

$$= \frac{12!}{2^6}\left[\left(\frac{1}{6}\right)\right]^6 = \frac{1,925}{559,872} = .0034$$

7.3   The Normal Distribution

The normal distribution is one of the most important examples of a continuous probability distribution. The equation

$$y = \frac{1}{\sigma\sqrt{2\pi}}\, e^{-\frac{(x-\mu)^2}{2\sigma^2}} \tag{5}$$

is called a normal curve or Gaussian distribution. In (5) μ denotes mean, and σ is the standard deviation. The total area bounded by (5) and the x-axis is one. Thus, the area bounded by curve (5), the x-axis, and $x = a$ and $x = b$, where $a < b$, represents the probability that $x \, \varepsilon \, [a, b]$, denoted by

$$p(a < x < b).$$

A new variable z can be introduced by

$$z = \frac{x - \mu}{\sigma}. \tag{6}$$

Then, equation (5) becomes

$$y = \frac{1}{\sqrt{2\pi}}\, e^{-\frac{1}{2}z^2} \tag{7}$$

where $\sigma = 1$. Equation (7) is called the standard form of a normal distribution. Here, z is normally distributed with a mean of zero and a variance of one. The graph of the standardized normal curve is shown on the following page.

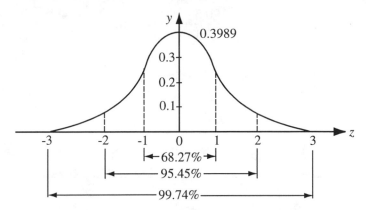

For example, the area between $z = -2$ and $z = 2$ under the curve is equal to 95.45% of the total area under the curve, which is 1.

We shall now list some properties of the normal distribution given by Equation (5).

Mean	μ
Standard deviation	σ
Variance	σ^2
Moment coefficient of skewness (curve is symmetric)	$a_3 = 0$
Moment coefficient of kurtosis	$a_4 = 3$

EXAMPLE:

The weights of 300 men were measured. The mean weight was 160 lbs., and the standard deviation was 14 lbs. Assume that weights are normally distributed. We shall determine

1. how many men weigh more than 190 lbs.

2. how many men weigh between 145 and 165 lbs.

1. The weight of each man was rounded to the nearest pound. Men weighing more than 190 lbs. must weigh at least 190.5 lbs. In standard units

$$\frac{190.5 - 160}{14} = 2.18.$$

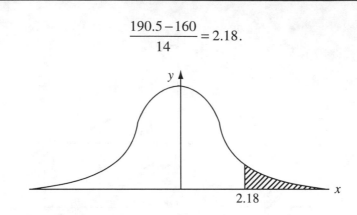

We have to find the shaded area.

Shaded Area = area to the right of 2.18

= (area to the right of $z = 0$) – (area between $z = 0$ and $z = 2.18$)

= 0.5 – 0.4854

= 0.0146

The number of men weighing more than 190 lbs. is

$$0.0146 \times 300 = 4.$$

2. Actually, we are interested in people weighing between 144.5 and 165.5 lbs. In standard units

$$\frac{144.5 - 160}{14} = -1.11, \quad \frac{165.5 - 160}{14} = 0.39.$$

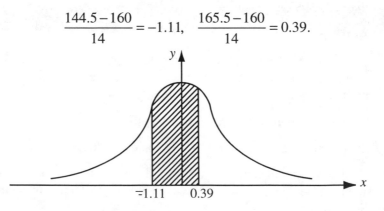

Area between $z = -1.11$ and $z = 0.39$

= (area between $z - 1.11$ and $z = 0$) + (area between $z = 0$ and $z = 0.39$)

= 0.3665 + 0.1517

= 0.5182.

The number of men weighing between 145 and 165 lbs. is

$$0.5182 \times 300 = 155.$$

Note that the areas of interest were obtained from the tables for the standard normal curve. In terms of probability, we can write the results

$$p\,(w \geq 190.5) = 0.0146$$

$$p\,(144.5 \leq w \leq 165.5) = 0.5182.$$

For number n large, and p and q, which are not too close to zero, the binomial distribution can be fairly well approximated by a normal distribution with the standardized variable

$$z = \frac{x - np}{\sqrt{npq}}.$$

With increasing n the approximation becomes better, and for $n \to \infty$ it becomes a normal distribution.

Problem Solving Examples:

 An electrical firm manufactures light bulbs which have a lifetime that is normally distributed with a mean of 800 hours and a standard deviation of 40 hours. Of 100 bulbs, about how many will have lifetimes between 778 and 834 hours?

 The probability that X, the lifetime of a randomly selected bulb, is between 778 and 834 hours is

$$Pr\,(778 < X < 834).$$

Standardizing, we see that $Pr\,(778 < X < 834)$

$$= \Pr\left(\frac{778-\mu}{\sigma} < \frac{x-\mu}{\sigma} < \frac{834-\mu}{\sigma}\right)$$

$$= \Pr\left(\frac{778-800}{40} < z < \frac{834-800}{40}\right)$$

$$= \Pr\left(\frac{-22}{40} < z < \frac{34}{40}\right)$$

$$= \Pr(-.55 < z < .85)$$

$$= \Pr(z < .85) - \Pr(z < -.55)$$

$$= .802 - (1 - .709)$$

$$= .511.$$

Thus, 51.1% of the bulbs manufactured will have lifetimes between 778 and 834 hours. On the average, 51 of 100 light bulbs will have lifetimes between 778 and 834 hours.

 A pair of dice is thrown 120 times. What is the approximate probability of throwing at least fifteen 7's? Assume that the rolls are independent, and remember that the probability of rolling a 7 on a single roll is $\frac{6}{36} = \frac{1}{6}$.

A The answer to this problem is a binomial probability. If $X =$ number of 7's rolled, and $n = 120$, then

$$\Pr(X \geq 15) = \sum_{j=15}^{120} \binom{120}{j}\left(\frac{1}{6}\right)^j \left(\frac{5}{6}\right)^{120-j}$$

This sum is quite difficult to calculate. There is an easier way. If n is large, $Pr_B (X \geq 15)$ can be approximated by $Pr_N (X \geq 14.5)$, where X is normally distributed with the same mean and variance as the binomial random variable. Remember that the mean of a binomially distributed random variable is np; n is the number of trials and p is the probability of "success" in a single trial. The variance of a binomially distributed random variable is $np(1-p)$ and the standard deviation is $\sqrt{np(1-p)}$.

Because of this fact $\dfrac{X - np}{\sqrt{np(1-p)}}$ is normally distributed with mean

0 and variance 1.

$$np = (120)\left(\frac{1}{6}\right) = 20,$$

and

$$\sqrt{np(1-p)} = \sqrt{120\left(\frac{1}{6}\right)\left(\frac{5}{6}\right)} = \sqrt{\frac{50}{3}} = 4.08248$$

Thus, $\quad \Pr(X \geq 15) = \Pr\left(\dfrac{X - 20}{4.08248} > \dfrac{14.5 - 20}{4.08248}\right)$

$$= \Pr(z > -1.35) = 1 - \Pr(Z < -1.35)$$

$$= 1 - .0885 = .9115$$

$$= \Pr_B(X > 15) \approx \Pr_N(X \geq 14.5)$$

The reason 15 has become 14.5 is that a discrete random variable is being approximated by a continuous random variable.

Consider the example below:

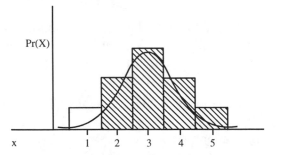

$Pr_B(2 \leq X \leq 5) =$ sum of the areas of the shaded rectangles. In approximating this area with a curve, we must start at the edge of the first shaded rectangle and move to the edge of the last shaded rectangle. This implies $Pr_B(2 \leq X \leq 5) \approx Pr_N(1.5 \leq X \leq 5.5)$.

7.4   The Poisson Distribution

The Poisson distribution is the discrete probability distribution defined by

$$p(x) = \frac{\lambda^x e^{-\lambda}}{x!} \tag{8}$$

where $x = 0, 1, 2, \ldots$ and e and λ are constants.

Properties of the Poisson Distribution

Mean	$\mu = \lambda$
Standard deviation	$\sigma = \sqrt{\lambda}$
Variance	$\sigma = \lambda$
Moment coefficient of skewness	$a_3 = \dfrac{1}{\sqrt{\lambda}}$
Moment coefficient of kurtosis	$a_4 = 3 + \dfrac{1}{\lambda}$

The Poisson distribution approaches a normal distribution with the standardized variable $\dfrac{x - \lambda}{\sqrt{\lambda}}$ as λ increases to infinity.

Under certain conditions, the binomial distribution is very closely approximated by the Poisson distribution. These conditions are that n is very large, the probability p of occurrence of an event is close to 0 (such an event is called a rare event), and $q = 1 - p$ is close to 1.

In applications, an event is considered rare if n is at least 50 ($n \geq 50$) while $np < 5$. Then the binomial distribution is closely approximated by the Poisson distribution with $\lambda = np$.

EXAMPLE:

Five percent of people have high blood pressure. Find the probability that in a sample of ten people chosen at random, exactly three will have high blood pressure.

First, we use the binomial distribution:

$$p(\text{high blood pressure}) = \frac{1}{20}$$

$$p(3 \text{ in } 10) =_{10} C_3 \left(\frac{1}{20}\right)^3 \left(1 - \frac{1}{20}\right)^7$$

$$= \frac{10!}{7!3!}\left(\frac{1}{20}\right)^3 \left(\frac{19}{20}\right)^7 = 0.01.$$

We solve this problem using the Poisson approximation to the binomial distribution

$$p(3 \text{ in } 10) = \frac{\lambda^x e^{-\lambda}}{x!}.$$

We set $\lambda = np = 10 \times \dfrac{1}{20} = 0.5$

$$p = \frac{(0.5)^3 e^{-0.5}}{3!} = 0.01.$$

In this case, $p = \dfrac{1}{20}$ (close to zero) and $np < 5$. Approximation is good.

Problem Solving Examples:

 Consider a production process of making ball bearings where the probability of a defective bearing is .01. What is the probability of having 10 defective bearings out of 1,000?

 Since n is very large and p is small, we can use the Poisson approximation to the binomial with $1 = E(x) = np = 1,000 \times .01 = 10$.

$$P(X = K) = \frac{e^{-10}(10)^k}{K!}$$

$$P(X = 10) = \frac{e^{-10}(10)^{10}}{10!} = .125$$

Let the probability of exactly one blemish in one foot of wire

be about $\dfrac{1}{1,000}$, and let the probability of two or more blem-

ishes in that length be, for all practical purposes, 0. Let the random variable X be the number of blemishes in 3,000 feet of wire. Find $Pr(X = 5)$.

 Let $n = 3,000$ feet of wire and $p = \dfrac{1}{1,000}$ of having a blemish

in one foot. We are dealing with a binomial random variable

with parameters $n = 3,000$ and $p = \dfrac{1}{1,000}$. The exact answer to this,

therefore, is

$$Pr(X = 5) = {}_{3,000}C_3 \left(\frac{1}{1,000}\right)^5 \left(\frac{999}{1,000}\right)^{2,995} = .1008.$$

This is an incredibly tedious computation. Instead, since n is large and p is small, we will use a Poisson approximation.

We know that λ must equal np.

$$\lambda = 3,000 \left(\frac{1}{1,000}\right) = 3.$$

Then

$$p(x = 5) = \frac{3^5 e^{-3}}{5!} = .101$$

Note that $Pr(X = 5) = Pr(X \le 5) - Pr(X \le 4)$.

The last two values can be read off the cumulative tables under expectation 3. Hence,

$$Pr(X = 5) = .916 - .815 = .101.$$

Quiz: Distributions

1. If a couple getting married today can be expected to have 0, 1, 2, 3, 4, or 5 children with probabilities of 20%, 20%, 30%, 20%, 8%, and 2%, respectively, what is the average number of children, to the nearest tenth, couples getting married today have?

   (A) 1.0   (D) 2.2

   (B) 1.8   (E) 2.8

   (C) 2.0

2. A well-balanced coin is flipped 10,000 times. What is the probability of getting more than 5,100 heads?

   (A) 42.17%   (D) 2.01%

   (B) 2.28%   (E) 2.00%

   (C) 2.22%

3. A woman has three pairs of socks similar in color. If she decides to pair the six socks in a random fashion, what is the probability that each sock has been properly matched?

   (A) $\dfrac{1}{15}$ (D) $\dfrac{1}{3}$

   (B) $\dfrac{1}{9}$ (E) $\dfrac{1}{2}$

   (C) $\dfrac{1}{6}$

4. Given a normal distribution (of continuous data) in which the mean is 150 and the standard deviation is 12, which score separates the lower 4% from the upper 96%?

   (A) 124   (D) 131

   (B) 126   (E) 134

   (C) 129

5. There are two finalists in a golf tournament, with the winner getting $40,000 and the runner-up getting $15,000. If the mathematical expectation of the better player is $32,500, what is the probability of winning?

   (A) .55   (D) .70

   (B) .60   (E) .75

   (C) .65

6. A doctor knows that 15% of all her patients are late for their appointments. Given five randomly selected patients, what is the approximate probability that exactly three of them are late for their appointments?

   (A) .004   (D) .114

   (B) .024   (E) .154

   (C) .064

7. Given events A and B where Prob $(A) = .25$, Prob $(B) = .30$; let Prob (A/B) mean the probability of event A given that event B has occurred. If Prob $(A/B) = .40$, what is the value of Prob (B/A)?

   (A) .45   (D) .62

   (B) .48   (E) .67

   (C) .54

8. Next weekend, you are expecting a visit from Bonnie and Clyde. The probability that Bonnie will show up is .4, whereas the probability that Clyde will show up is .8. What is the probability that at least one of them will visit you next weekend?

   (A) .32   (D) .76

   (B) .40   (E) .88

   (C) .52

9. Using the Poisson distribution function, if 1.2 accidents can be expected at a certain intersection every day, what is the approximate probability that there will be two accidents at that intersection on any given day?

   (A) .34

   (B) .30

   (C) .26

   (D) .22

   (E) .18

10. Mark, Linda, Bill, and Joan are billing clerks in an office. Of the number of erroneous billings prepared, 40% were done by Mark, 20% were done by Bill, 10% were done by Linda, and the rest were done by Joan. Given seven random erroneous billings, what is the approximate probability that two were prepared by Mark, one by Bill, one by Linda, and three by Joan?

    (A) .036

    (B) .047

    (C) .058

    (D) .069

    (E) .082

ANSWER KEY

1.	(B)	6.	(B)
2.	(C)	7.	(B)
3.	(A)	8.	(E)
4.	(C)	9.	(D)
5.	(D)	10.	(A)

CHAPTER 8

Sampling Theory

8.1 Sampling

For a statistician, the relationship between samples and population is important. This branch of statistics is called **sampling theory**. We gather all pertinent information concerning the sample in order to make statements about the whole population.

Sample quantities, such as sample mean, deviation, etc., are called **sample statistics** or **statistics**. Based on these quantities, we estimate the corresponding quantities for population, which are called **population parameters** or **parameters**. For two different samples, the difference between sample statistics can be due to chance variation or some significant factor. The latter case should be investigated, and possible mistakes corrected. The statistical inference is a study of inferences made concerning a population and based on the samples drawn from it.

Probability theory evaluates the accuracy of such inferences. The most important initial step is the choice of samples that are representative of a population. The methods of sampling are called the **design** of the experiment. One of the most widely used methods is random sampling.

8.1.1 Random Sampling

A sample of n measurements chosen from a population N ($N > n$)

is said to be a random sample if every different sample of the same size *n* from the population has an equal probability of being selected.

One way of obtaining a random sample is to assign a number to each member of the population. The population becomes a set of numbers. Then, using the random number table, we can choose a sample of desired size.

EXAMPLE:

Suppose 1,000 voters are registered and eligible to vote in an upcoming election. To conduct a poll, you need a sample of 50 persons, so to each voter you assign a number between one and 1,000. Then, using the random number table or a computer program, you choose at random 50 numbers, which are 50 voters. This is your required sample.

8.1.2 Sampling With and Without Replacement

From a bag containing ten numbers from 1 to 10, we have to draw three numbers. As the first step, we draw a number. Now, we have the choice of replacing or not replacing the number in the bag. If we replace the number, then this number can come up again. If the number is not replaced, then it can come up only once.

Sampling where each element of a population may be chosen more than once (i.e., where the chosen element is replaced) is called **sampling with replacement**. Sampling without replacement takes place when each element of a population can be chosen only once.

Remember that populations can be finite or infinite.

EXAMPLE:

A bag contains ten numbers. We choose two numbers without replacement. This is sampling from a finite population.

EXAMPLE:

A coin is tossed ten times and the number of tails is counted. This is sampling from an infinite population.

Problem Solving Examples:

 The following sampling procedure is to be classified as producing a random sample or as producing a biased sample. Decide whether the procedure is random or biased.

In order to solve a particular problem, an investigator selects 100
people, each of which will provide 5 scores. The investigator will then
take the average of each set of scores. This will yield 100 averages. Is
this sample of average scores a random sample?

 A sample must meet the following conditions in order to be
random:

(1) Equal Chance. A sample meets the condition of equal chance if
it is selected in such a way that every observation in the entire population
has an equal chance of being included in the sample.

(2) Independence. A sample meets this condition when the selection
of any single observation does not affect the chances for selection of any
other.

Samples that are not random are called biased.

The 100 elements are now independent. When repeated measures
can be converted to a single score, so that each individual observed
contributes just one summary observation (such as an average), the
independence condition is met. Use of an average often helps to reduce
the effects of chance variation within an individual's performance.
Also, any observation would have an equal chance of being chosen. The
sample is random.

A wheat researcher is studying the yield of a certain variety of
wheat in the state of Colorado. She has at her disposal five
farms scattered throughout the state on which she can plant the wheat
and observe the yield. Describe the sample and the target population.
Under what conditions will this be a random sample?

The sample consists of the wheat yields on the five farms.
The target population consists of the yields of wheat on every

farm in the state. This sample will be random if (1) every farm in the state has an equal chance of being selected and (2) the selection of any particular farm is independent of the selection of any other farm.

8.2 Sampling Distributions

A population is given from which we draw samples of size n, with or without replacement. For each sample, we compute a statistic, such as the mean, standard deviation, variance, etc. These numbers will depend on the sample, and they will vary from sample to sample. In this way, we obtain a distribution of the statistic which is called the **sampling distribution**.

For example, if for each sample we measure its mean, then the distribution obtained is the sampling distribution of means. We obtain the sampling distributions of variances, standard deviations, medians, etc. in the same way.

8.3 The Central Limit Theorem

A population is given with a finite mean, μ, and a standard deviation, σ. Random samples of n measurements are drawn. If the population is infinite, or if sampling is with replacement, then the relative frequency histogram for the sample means will be approximately normal with mean μ and standard deviation $\dfrac{\sigma}{\sqrt{n}}$.

Suppose the distribution of x for the population is as shown in the figure below, with the mean μ.

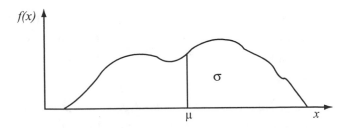

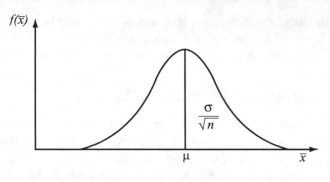

The standard deviation is σ.

The figure above shows the relative frequency distribution, called the sampling distribution, for the sample mean $\bar{x}$. The samples with replacement of size n are measured and sample mean $\bar{x}$ is computed. The mean for the sampling distribution of $\bar{x}$ is μ, the same as for the whole population. The standard deviation of the sampling distribution is equal to the standard deviation of the x measurements divided by $\sqrt{n}$, that is, $\dfrac{\sigma}{\sqrt{n}}$.

If the samples of size n are drawn without replacement from a finite population of size N, then

$$\mu_{\bar{x}} = \mu \ \text{ and } \ \sigma_{\bar{x}} = \frac{\sigma}{\sqrt{n}} \times \sqrt{\frac{N-n}{N-1}}$$

where μ and σ denote the population mean and standard deviation; while $\mu_{\bar{x}}$ and $\sigma_{\bar{x}}$ denote the mean and standard deviation, respectively, of the sampling distribution.

EXAMPLE:

Suppose a population consists of the five numbers 2, 4, 5, 6, and 8. All possible samples of size two are drawn with replacement. Thus, there are $5 \times 5 = 25$ samples. The mean of the population is

$$\mu = \frac{25}{5} = 5$$

and the standard deviation of the population is

$$\sigma^2 = \frac{(2-5)^2 + (4-5)^2 + (5-5)^2 + (6-5)^2 + (8-5)^2}{5} = 4$$

$$\sigma = 2$$

We shall list all 25 samples and their corresponding sample means

(2, 2); 2	(4, 2); 3	(5, 2); 3.5	(6, 2); 4	(8, 2); 5
(2, 4); 3	(4, 4); 4	(5, 4); 4.5	(6, 4); 5	(8, 4); 6
(2, 5); 3.5	(4, 5); 4.5	(5, 5); 5	(6, 5); 5.5	(8, 5); 6.5
(2, 6); 4	(4, 6); 5	(5, 6); 5.5	(6, 6); 6	(8, 6); 7
(2, 8); 5	(4, 8); 6	(5, 8); 6.5	(6, 8); 7	(8, 8); 8

The mean of the sampling distribution of means is

$$\mu_{\bar{x}} = \frac{\text{sum of means}}{25} = \frac{125}{25} = 5.$$

Thus,

$$\mu = \mu_{\bar{x}}.$$

The variance $\sigma_{\bar{x}}^2$ is

$$\sigma_{\bar{x}}^2 = \frac{\sum (\mu_n - \mu_{\bar{x}})}{25} = \frac{50}{25} = 2$$

and $\sigma_{\bar{x}} = 1.414$.

For finite populations and sampling with replacement (or infinite populations), we have

$$\sigma_{\bar{x}} = \frac{\sigma}{\sqrt{n}} = \frac{2}{\sqrt{2}} = 1.414$$

which agrees with the calculated value.

Problem Solving Examples:

 A population consists of the number of defective transistors in shipments received by an assembly plant. The number of defectives is 2 in the first, 4 in the second, 6 in the third, and 8 in the fourth.

(a) Find the mean μ and the standard deviation σ of the given population.

(b) List all random samples, with replacement, of size 2 that can be formed from the population and find the distribution of all sample means.

(c) Find the mean and the standard deviation of the sampling distribution of the mean.

 The population is 2, 4, 6, 8.

(a) $\mu = \dfrac{\sum_{i=1}^{n} x_i}{n} = \dfrac{\sum_{i=1}^{4} x_i}{4} = \dfrac{2+4+6+8}{4} = \dfrac{20}{4} = 5$

We will compute

$$\sigma = \frac{\sum_{i=1}^{n}(x_i - \mu_{\bar{x}})^2}{n} = \sqrt{\frac{(2-5)^2 + (4-5)^2 + (6-5)^2 + (8-5)^2}{4}}$$

$$= \sqrt{\frac{9+1+1+9}{4}} = \sqrt{\frac{20}{4}} = \sqrt{5}.$$

(b) The following table should prove useful.

	Sample	Sample Mean
1.	2, 2	2
2.	2, 4	3
3.	2, 6	4
4.	2, 8	5
5.	4, 2	3
6.	4, 4	4
7.	4, 6	5
8.	4, 8	6
9.	6, 2	4
10.	6, 4	5
11.	6, 6	6
12.	6, 8	7
13.	8, 2	5
14.	8, 4	6
15.	8, 6	7
16.	8, 8	8

Collating the data we have

x = Sample Mean	$N(x)$ = Number of times x occurs	$F(x) = N(x)/n$ $= N(x)/16$
2	1	1/16
3	2	1/8
4	3	3/16
5	4	1/4
6	3	3/16
7	2	1/8
8	1	1/16

(c) Sample Mean $= \dfrac{\displaystyle\sum_{i=1}^{n} i^{th} \text{ Sample Mean}}{n}$

$= \dfrac{2+3+4+5+3+4+5+6+4+5+6+7+5+6+7+8}{16}$

$= \dfrac{80}{16} = 5.$

$\sigma_{\bar{x}} = \sqrt{\dfrac{\sum(\mu - \mu_{\bar{x}})}{n}}$

$= \sqrt{\dfrac{(2-5)^2 + 2(3-5)^2 + 3(4-5)^2 + 4(5-5)^2 + 3(6-5)^2 + 2(7-5)^2 + (8-5)^2}{16}}$

$= \sqrt{\dfrac{9+8+3+0+3+8+9}{16}} = \sqrt{\dfrac{40}{16}} = \sqrt{\dfrac{5}{2}}.$

A population consists of 1, 4, and 10. Find μ, σ, $\mu_{\bar{x}}$, and $\sigma_{\bar{x}}$. List all samples of size 2, without replacement.

$\mathbf{A}$ $\mu = \dfrac{\displaystyle\sum_{i=1}^{3} X_i}{N} = 3$

$\sigma = \sqrt{\dfrac{\displaystyle\sum_{i=1}^{n}(X_i - \mu)^2}{3}}$

$\mu = \dfrac{(1+4+10)}{3} = 5$

$\sigma = \sqrt{\dfrac{\left[(1-5)^2 + (4-5)^2 + (10-5)^2\right]}{3}}$

$= \sqrt{\dfrac{42}{3}} = \sqrt{14}$

$\mu_{\bar{x}} = \mu = 5$

$$\sigma_{\bar{x}} = \left(\frac{\sqrt{14}}{\sqrt{2}} \right)\left(\sqrt{\frac{(3-2)}{(3-1)}} \right)$$

$$= \left(\sqrt{7} \right)\left(\sqrt{\frac{1}{2}} \right) = \sqrt{3.5} \approx 1.87$$

The three samples are 1, 4; 1, 10; and 4, 10; their means are 2.5, 5.5, and 7. Then $\mu_{\bar{x}} = (2.5 + 5.5 + 7) \div 3 = 5$. Now $\sigma_{\bar{x}} = \dfrac{\sigma}{\sqrt{n}} \sqrt{\dfrac{N-n}{N-1}}$, where $n = 2$, the size of each sample.

8.4 Sampling Distribution of Proportions

Suppose that for an infinite population the probability of occurrence of an event (i.e., its success) is p. Then, the probability of its failure is $q = 1 - p$.

EXAMPLE:

The population is all possible tosses of a coin (we always assume a coin, a die, etc., to be fair). The probability of the event "tails" is $p = \dfrac{1}{2}$.

All possible samples of size n are drawn from this population and for each sample the proportion P of success is determined. For n tosses of the coin, P is the proportion of tails obtained. We have a sampling distribution of proportions. Its mean, μ_P, and standard deviation, σ_P, are given by

$$\mu_P = p$$

$$\sigma_P = \sqrt{\frac{p(1-p)}{n}} = \sqrt{\frac{pq}{n}}$$

The equations are valid for a finite population and sampling with replacement. For large values of n, the sampling distribution is close to the normal distribution. For finite populations and sampling without replacement, the mean, μ_p, is

$$\mu_P = p$$

and the standard deviation, σ_p, is

$$\sigma_P = \sqrt{p(1-p)}$$

Problem Solving Examples:

A fair coin is tossed six times. Find the mean and standard deviation for the number of tails.

$\mu_p = p$ = probability of success on one toss.

$$\sigma_P = \sqrt{\frac{pq}{n}}$$

p = probability of success
q = probability of failure
n = number of tosses

$$\mu_P = p = \frac{1}{2}$$

$$\sigma_P = \sqrt{\frac{\left(\frac{1}{2}\right)\left(1-\frac{1}{2}\right)}{6}} = \sqrt{\frac{1}{24}} \text{ or } \frac{1}{12}\sqrt{6}$$

$$\approx .204.$$

A fair die is tossed four times. Find the mean and standard deviation for the number of times 2 appears on the die.

$\mu_p = \dfrac{1}{6}$

$$\sigma_P = \sqrt{\frac{\left(\frac{1}{6}\right)\left(1-\frac{1}{6}\right)}{4}} = \sqrt{\frac{5}{144}}\frac{1}{12}\sqrt{5}$$

$$\approx 0.186.$$

8.5 Sampling Distributions of Sums and Differences

Two populations are given. From the first population, we draw samples of size n_1 and compute a statistic, s_1. Thus, we obtain a sampling distribution for the statistic, s_1, with the mean, μ_{s_1}, and standard deviation, σ_{s_1}.

Similarly, from the second population we draw samples of size n_2, compute a statistic s_2, and find μ_{s_2} and σ_{s_2}. From all possible combinations of these samples from the two populations, we can determine a distribution of the sums, $s_1 + s_2$, which is called the sampling distribution of sums of the statistics. We can also find a distribution of the differences, $s_1 - s_2$, which is called the sampling distribution of differences of the statistics. The mean and standard deviation of the sampling distribution is

for the sum

$$\mu_{s_1+s_2} = \mu_{s_1} + \mu_{s_2}$$

$$\sigma_{s_1+s_2} = \sqrt{\sigma_{s_1}^2 + \sigma_{s_2}^2}$$

for the difference

$$\mu_{s_1-s_2} = \mu_{s_1} - \mu_{s_2}$$

$$\sigma_{s_1-s_2} = \sqrt{\sigma_{s_1}^2 + \sigma_{s_2}^2}$$

The samples have to be independent; that is, they do not depend upon each other.

EXAMPLE:

Suppose f_1 can be any of the elements of the population 3, 5, 7, and suppose that f_2 can be any of the elements of the population 2, 4. Then

$$\mu_{f_1} = \text{mean of population } f_1 = \frac{3+5+7}{3} = 5$$

$$\mu_{f_2} = \text{mean of population } f_2 = \frac{2+4}{2} = 3$$

Now, let us consider the population consisting of the sums of any number of f_1 and any number of f_2.

$$3 + 2 \quad 5 + 2 \quad 7 + 2$$
$$3 + 4 \quad 5 + 4 \quad 7 + 4$$

or

$$5 \quad 7 \quad 9$$
$$7 \quad 9 \quad 11$$

The mean $\mu_{f_1+f_2}$ is

$$\mu_{f_1+f_2} = \frac{48}{6} = 8.$$

This result is in agreement with the general rule

$$8 = \mu_{f_1+f_2} = \mu_{f_1} + \mu_{f_2} = 5 + 3.$$

The standard deviations are

$$\sigma_{f_1}^2 = \frac{(3-5)^2 + (5-5)^2 + (7-5)^2}{3} = 2.667$$

$$\sigma_{f_1} = 1.633 \text{ and}$$

$$\sigma_{f_2}^2 = \frac{(2-3)^2 + (4-3)^2}{2} = 1 \quad \sigma_{f_2} = 1.$$

Similarly, we compute $\sigma_{f_1 + f_2}$

$$\sigma^2_{f_1+f_2} = \frac{(5-8)^2+(7-8)^2+(9-8)^2+(7-8)^2+(9-8)^2+(11-8)^2}{6}$$

$$= 3.667$$

$$\sigma_{f_1+f_2} = 1.915.$$

This agrees with the general formula

$$\sigma_{f_1+f_2} = \sqrt{\sigma^2_{f_1} + \sigma^2_{f_2}}$$

for independent samples.

Suppose s_1 and s_2 are the sample means from the two populations, which we denote by $\bar{x}_1$ and $\bar{x}_2$. Then, for infinite populations (or finite populations and sampling with replacement) with means μ_1 and μ_2 and standard deviations σ_1 and σ_2, respectively, the sampling distribution of the sums (or the differences) of means is for sums

$$\mu_{\bar{x}_1+\bar{x}_2} = \mu_{\bar{x}_1} + \mu_{\bar{x}_2} = \mu_1 + \mu_2$$

$$\sigma_{\bar{x}_1+\bar{x}_2} = \sqrt{\sigma^2_{\bar{x}_1} + \sigma^2_{\bar{x}_2}} = \sqrt{\frac{\sigma_1^2}{n_1} + \frac{\sigma_2^2}{n_2}}$$

and for differences

$$\mu_{\bar{x}_1-\bar{x}_2} = \mu_{\bar{x}_1} - \mu_{\bar{x}_2}$$

$$\sigma_{\bar{x}_1 - \bar{x}_2} = \sqrt{\sigma_{\bar{x}_1}^2 + \sigma_{\bar{x}_2}^2} = \sqrt{\frac{\sigma_1^2}{n_1} + \frac{\sigma_2^2}{n_2}}$$

where n_1 and n_2 are sizes of samples.

EXAMPLE:

Two producers manufacture tires. The mean lifetime of tires made by A is 120,000 miles with a standard deviation of 20,000 miles, while the mean lifetime of tires made by B is 80,000 miles with a standard deviation of 10,000 miles.

Random samples of 200 tires of each brand are tested. What is the probability that tires made by A will have a mean lifetime which is at least 45,000 miles more than the tires made by B? $\bar{x}_A$ and $\bar{x}_B$ denote the mean lifetimes of samples A and B, respectively. Then

$$\mu_{\bar{x}_A - \bar{x}_B} = \mu_{\bar{x}_A} - \mu_{\bar{x}_B} = 120,000 - 80,000 = 40,000 \text{ miles}$$

$$\sigma_{\bar{x}_A - \bar{x}_B} = \sqrt{\frac{\sigma_A^2}{n_A} + \frac{\sigma_B^2}{n_B}} = \sqrt{\frac{(20,000)^2}{200} + \frac{(10,000)^2}{200}}$$

$$= 1581$$

The standardized variable for the difference in means is

$$z = \frac{(\bar{x}_A - \bar{x}_B) - (\mu_{\bar{x}_A - \bar{x}_B})}{\sigma_{\bar{x}_A - \bar{x}_B}} = \frac{(\bar{x}_A - \bar{x}_B) - 40,000}{1581}$$

For large samples, the distribution is normal. For the difference 45,000 miles

$$z = \frac{45,000 - 40,000}{1,581} = 3.16.$$

Hence,

required probability =
area under normal curve to right of $z = 3.16 = 0.5 - 0.4992 = 0.0008$

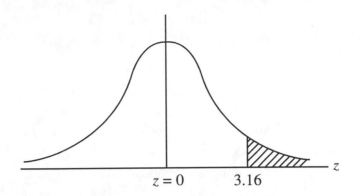

$$z = 0 \qquad 3.16$$

Problem Solving Examples:

Q Population A has a mean of 5 and a standard deviation of 3, whereas population B has a mean of 12 and a standard deviation of 2. Find the mean and standard deviation of the sampling distribution of sum of means. Samples of size 4 are selected from A, whereas samples of size 6 are selected from B. Assume A and B are infinite in size.

A $\mu_{\overline{X}_A + \overline{X}_B}$ means: the mean of the sampling distribution of the mean of a sample from A combined with a mean of a sample from B. Essentially, all possible pairings of numbers from A and B are made, the mean of each pairing calculated, then the overall mean of these means is computed.

$$\mu_{\overline{X}_A + \overline{X}_B} = 5 + 12 = 17$$

$\sigma_{\overline{X}_A + \overline{X}_B}$ is the standard deviation for the distribution described above.

$$\sigma_{\overline{X}_A + \overline{X}_B} = \sqrt{\frac{\sigma_A^2}{n_A} + \frac{\sigma_B^2}{n_B}} = \sqrt{\frac{9}{4} + \frac{4}{6}} = \sqrt{\frac{35}{12}}$$

$$= \sqrt{2.917} \approx 1.708$$

Using the information provided in the previous question, find the mean and standard deviation for the sampling distribution of the difference of means.

$\mu_{\overline{x}_A - \overline{x}_B}$ = mean of sampling distribution of differences of means from A to B. A value from A is chosen, then subtracted by a value from B. All possible pairings are done to create a distribution of differences.

$\sigma_{\overline{x}_A - \overline{x}_B}$ = standard deviation of this distribution.

$$\mu_{\overline{x}_A - \overline{x}_B} = 5 - 12 = -7$$

$$\sigma_{\overline{x}_A - \overline{x}_B} = \sqrt{\frac{\sigma_A^2}{n_A} + \frac{\sigma_B^2}{n_B}} = \sqrt{\frac{35}{12}}$$

$$= \sqrt{2.917} \approx 1.708$$

8.6   Standard Errors

The standard deviation of a sampling distribution of a statistic is called its **standard error**. If the sample size n is large enough, then the sampling distributions are nearly normal. Statistical methods dealing with such samples are called **large sampling methods**. When $n < 30$, the samples are small so the theory of small samples, or exact sampling theory, is applied. We shall denote

	Population	Sample
Mean	μ	$\overline{x}$
Standard Deviation	σ	s
Proportion	p	P
rth Moment	μ_r	m_r

If the population parameters are unknown, we can obtain close estimates from their corresponding sample statistics, assuming that the samples are large enough.

CHAPTER 9

Statistical Inference:
Large Samples

9.1 Estimation of Parameters

In most statistical applications, we move from a sample to the population. First, we draw a sample from its population in accordance with sampling theory. Then, all sample statistics (like sample mean, standard deviation, etc.) are measured to estimate population parameters (like population mean, standard deviation, etc.).

Statistical inference deals with the estimation of population parameters from the corresponding sample statistics.

In general, all inferences belong to one of the two categories.

EXAMPLE:

A new law is about to be introduced in the State Senate, and questionnaires are given to a sample of residents asking them for possible amendments to this law. Since the law is new, no information is available concerning the attitudes in the past. Thus, it is impossible to estimate the average response. But, having the sample information, we can estimate the response of all residents.

EXAMPLE:

A manufacturer tests a new electric motor for its washing machine. On the average, the old one could work 7,000 hours without repair, $\mu = 7,000$. If the new motor is better than the old one, it should, on the average, work longer than 7,000 hours without repair. Thus, we are testing a hypothesis that for a new motor $\mu > 7,000$. These two examples illustrate the two different inference-making procedures used:

1. estimation

2. test of hypothesis

9.1.1 Unbiased and Biased Estimates

If, for the sampling distribution, the mean of a statistic is equal to the corresponding population parameter, then the statistic is called an unbiased estimator of the parameter; otherwise, it is called a biased estimator.

EXAMPLE:

The sample mean, $\bar{x}$, is an unbiased estimate of the population mean μ. This is a direct result of the fact that

$$\mu_{\bar{x}} = \mu.$$

EXAMPLE:

The sample variance s^2 is a biased estimate of the population variance σ^2. The mean of the sampling distribution of variance is given by

$$\mu_{s^2} = \frac{n-1}{n}\,\sigma^2$$

where σ^2 is the population variance and n is the sample size.

Problem Solving Examples:

 A population has a variance of 10. All samples of size 5 are selected from this infinite population. Find the mean of the sampling distribution of the variance.

μ_{s2} = mean of the sampling distribution of variance

$$= \frac{(n-1)\sigma^2}{n}$$

$$= \frac{(4)(10)}{5} = 8.$$

Q With a population (infinite) that has a variance of 25, all samples of a specific size are selected. If the mean of the sampling distribution of the variance is $24\frac{1}{6}$, what is the size of each sample?

 $\mu_{s2} = \mu_{s^2} = \dfrac{(n-1)\sigma^2}{n}$,

so we have

$$24\frac{1}{6} = \frac{(n-1)(25)}{n}$$

Multiply both sides by n to get

$$24\frac{1}{6}n = (n-1)(25)$$

$$24\frac{1}{6}n = 25n - 25$$

$$-\frac{5}{6}n = -25$$

Solving, $n = 30$.

9.2 Point Estimation of μ

An estimate of a population parameter given by a single number is called a point estimate of the parameter. An estimate of a population parameter given by two numbers between which the parameter lies is called an interval estimate of the parameter.

EXAMPLE:

The distance is given as 3.45 miles. This is a point estimate. If the distance is given as 3.45 ± 0.02 miles, then we have an interval estimate. Here, the distance and the accuracy of the measurement are given. The distance is between 3.43 and 3.47 miles.

In statistical inference-making procedures, we not only make an inference but also supply information on how good the inference is. For example, from the measurements of a sample we obtain the value of $\bar{x}$ which is a point estimate of μ. But, there are many possible values of $\bar{x}$ and it may happen that our $\bar{x}$ will not be exactly equal to μ. Thus, we have to supply another piece of information, which is how close $\bar{x}$ is to the parameter we are trying to estimate. The Central Limit Theorem for $\bar{x}$ states that under certain conditions the sampling distribution of $\bar{x}$ would be approximately normal, with mean μ the same as for the population and standard deviation $\dfrac{\sigma}{\sqrt{n}}$, where σ is a standard deviation of a population and n is the size of samples. For the Central Limit Theorem to hold, the sample size must be at least 30.

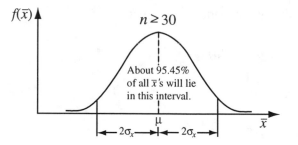

Having the sampling distribution of $\bar{x}$ and areas under the normal curve, we know that about 95.45% of the point estimates calculated from repeated samples will be within

$$2\sigma_{\bar{x}} = 2\frac{\sigma}{\sqrt{n}} \text{ of } \mu$$

as shown in the figure.

This information tells us how good our point estimate is.

9.2.1 Error of Estimation

For a given point estimate, the error of estimation is defined to be

$$\text{Error of estimation} = |\bar{x} - \mu|$$

9.2.2 Bound on Error

By the Empirical Rule, the error of estimation will be less than $2\sigma_{\bar{x}}$ approximately 95.45% of the time. The quantity

$$2\sigma_{\bar{x}}$$

is called a bound on the error of estimation. The smaller the bound on the error of estimation, the better the inference.

Point Estimation of a Population Mean μ

Point estimate of μ	$\bar{x}$
Sample size n	$n \geq 30$
Bound on the error	$2\sigma_{\bar{x}} = 2\dfrac{\sigma}{\sqrt{n}}$

EXAMPLE:

A car dealer, who sells new cars, is interested in estimating the average number of years a new car can run without any repairs. A random sample of $n = 250$ cars yields the average 2.6 years with a standard deviation of 1.2 years. We have

$$\bar{x} = 2.6 \quad s = 1.2 \quad n = 250$$

Hence, our point estimate is $\mu = 2.6$, and the corresponding bound on the error of estimation is given by

$$\text{bound on error} = 2\sigma_{\bar{x}} = \frac{2\sigma}{\sqrt{n}}$$

In this case, σ is unknown, but $n \geq 30$ and we can replace σ with s and get an approximate bound on the error

$$2\sigma_{\bar{x}} \approx \frac{2s}{\sqrt{n}} = \frac{2 \times 1.2}{\sqrt{250}} = 0.152.$$

Ninety-five percent of the sample means calculated in repeated sampling would be within two standard deviations. We can be fairly certain that

$$\bar{x} = 2.6$$

is within 0.152 of the actual mean μ.

Problem Solving Example:

 A random sample of 300 college students yields a grade point average of 2.8 with a standard deviation of .1.

(a) If, about 95.45% of the time, 2.8 will be in error by less than $2\sigma_{\bar{x}}$, what is the value of $2\sigma_{\bar{x}}$?

(b) What would have been the value of $2\sigma_{\bar{x}}$ if 400 students were selected?

 (a) $2\sigma_{\bar{x}} = \frac{2s}{\sqrt{n}} = \frac{(2)(.1)}{\sqrt{300}} \approx .012.$

$\sigma_{\bar{x}}$ = standard deviation of the sampling distribution of means. If all possible groups of 300 students were used (an impossible task), we could be 95.45% certain that the grand mean of all these groups would lie within .012 of the number 2.8. This means $2.8 \pm .012$, that is between 2.788 and 2.812.

(b) $2\sigma_{\bar{x}} = \frac{2s}{\sqrt{n}} = \frac{(2)(.1)}{\sqrt{400}} = .01.$

9.3   Interval Estimation of μ

Having a point estimate $\bar{x}$, we can obtain an interval estimate for the population mean μ.

The interval $[\mu - 1.96\ \sigma_{\bar{x}}, \mu + 1.96\ \sigma_{\bar{x}}]$ includes 95% of the x's in repeated sampling. See Figure 1.

Now consider the interval $[\bar{x} - 1.96\ \sigma_{\bar{x}}, \bar{x} + 1.96\ \sigma_{\bar{x}}]$. For simplicity, interval $[x - a, x + a]$ will be denoted by $x \pm a$. Whenever $\bar{x} \varepsilon \mu \pm 1.96\ \sigma_{\bar{x}}$, the intervals $\bar{x} \pm 1.96\ \sigma_{\bar{x}}$ will contain the parameter μ. This happens 95% of the time. See Figure 2.

The interval

$$\bar{x} \pm 1.96\ \sigma_{\bar{x}}$$

represents an interval estimate of μ.

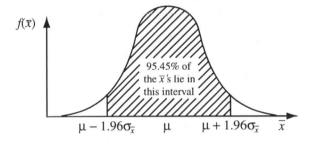

Figure 1

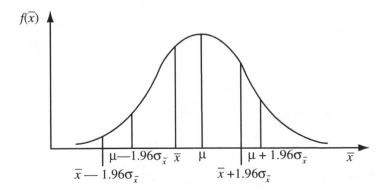

Figure 2

If $\bar{x}\,\varepsilon\,\mu\pm1.96\,\sigma_{\bar{x}}$, then $\mu\,\varepsilon\,\bar{x}\pm1.96\,\sigma_{\bar{x}}$. We measure $\bar{x}$ of a sample and find an interval $\bar{x}\pm1.96\,\sigma_{\bar{x}}$. The probability that this interval contains μ is 0.95.

9.3.1 Confidence Coefficient

The decimal 0.95 for the interval $\bar{x}\pm1.96\,\sigma_{\bar{x}}$ is called the confidence coefficient. In repeated sampling, 95% of the time the intervals calculated $\bar{x}\pm1.96\,\sigma_{\bar{x}}$ will contain the mean μ.

In practice, we calculate $\bar{x}$ and only one interval $\bar{x}\pm1.96\,\sigma_{\bar{x}}$. This interval, called a 95% confidence interval, represents an interval estimate of μ. Of course, one can construct many different confidence intervals for μ, depending on the chosen confidence coefficient.

The interval

$$\mu\pm2.58\,\sigma_{\bar{x}}$$

would include 99% of the values of $\bar{x}$ in repeated sampling.

The interval

$$\bar{x}\pm2.58\,\sigma_{\bar{x}}$$

forms a 99% confidence interval for μ.

EXAMPLE:

For a random sample of $n=40$ students, the average weight was 172 lbs. with a standard deviation of 54 lbs. Thus,

$$\bar{x}=172 \quad s=54 \quad n=40$$

The 95% confidence interval is given by

$$\bar{x}\pm1.96\,\sigma_{\bar{x}}$$

where $\sigma_{\bar{x}}=\dfrac{\sigma}{\sqrt{n}}$.

We shall substitute the sample standard deviation s for σ. Hence, the 95% confidence interval is

$$172 \pm \frac{1.96 \times 54}{\sqrt{40}} = 172 \pm 16.7$$

and the 99% confidence interval is

$$172 \pm \frac{2.58 \times 54}{\sqrt{40}} = 172 \pm 22$$

We define

$$\text{Confidence coefficient} = 1 - \alpha$$

where $0 \leq \alpha \leq 1$; α = level of significance.

For a specific value of $1 - \alpha$, a $100\,(1 - \alpha)\%$ confidence interval for μ is given by

$$\bar{x} \pm z_{\frac{\alpha}{2}} \sigma_{\bar{x}}$$

The value of $z_{\frac{\alpha}{2}}$ is such that at a distance of $z_{\frac{\alpha}{2}}$ standard deviations to the right of μ, the area under the normal curve is $\dfrac{\alpha}{2}$.

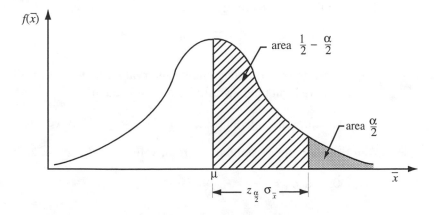

The table below gives the confidence coefficient and the corresponding values of $z_{\frac{\alpha}{2}}$.

Confidence Coefficient $1 - \alpha$	$z_{\frac{\alpha}{2}}$
0.9973	3
0.99	2.58
0.98	2.33
0.96	2.05
0.95	1.96
0.90	1.645
0.80	1.28
0.683	1
0.5	0.6745

Problem Solving Example:

In a random selection of 49 light bulbs from a specific manufacturer, the average lifetime of each bulb was 1,000 hours, with a standard deviation of four hours.

(a) Find the 95% confidence interval for the mean lifetime of all light bulbs from this manufacturer.

(b) What is the 99% confidence interval?

(a) We are trying to determine an interval so that 95% of all groups of 49 light bulbs would have an average lifetime within

that interval. The interval is given by $\bar{x} \pm Z_c \dfrac{\sigma}{\sqrt{n}}$, where Z_c is a critical

value taken from a table of standard normal values. In this case, $Z_c =$

1.96, $x = 1,000$, and $\sigma_{\bar{x}} = \dfrac{\sigma}{\sqrt{n}} = \dfrac{4}{\sqrt{49}} = \dfrac{4}{7}$.

Then, $1,000 \pm (1.96) \left(\dfrac{4}{7} \right) = 1,000 \pm 1.12$ is the 95% confidence interval.

(b) $1,000 \pm (2.58) \left(\dfrac{4}{7} \right) \approx 1,000 \pm 1.474$.

9.4 Test of Hypothesis

We shall consider a statistical test of a hypothesis. This test should answer the question, "Is the population mean equal to a specified value μ_0?" A statistical test consists of the following parts:

1. Research hypothesis, denoted by h_1
2. Null hypothesis, denoted by h_0
3. Test statistic
4. Rejection (or acceptance) region
5. Conclusion

For example, we are interested in what profit a $1,000 deposit would yield in different banks.

The research hypothesis is that the mean yield per $1,000 is greater than $125, which is the average observed for banks in the past three years. We formulate the null hypothesis, that

$$\mu = 125$$

Research hypothesis $\qquad\qquad \mu > 125$

Null hypothesis $\qquad\qquad \mu \leq 125$

To verify the research hypothesis, we must contradict the null hypothesis. A random sample of deposits gives $\bar{x}$ and s, that is the sample mean and standard deviation, respectively.

Whether we accept or reject the null hypothesis is based on a test statistic computed from the sample data. The value $\bar{x}$ can be chosen as the test statistic. Assuming that the null hypothesis is true, the sampling distribution of $\bar{x}$ is approximately normal with mean μ. The values $\bar{x}$

located in the right slope of the distribution will be contradictory to the null hypothesis and in favor of the research hypothesis. These values form a rejection region for our statistical test.

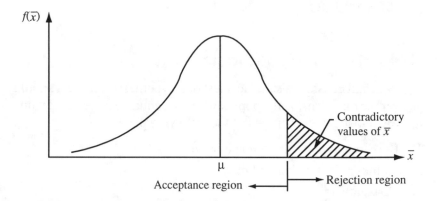

If the observed value of $\bar{x}$ falls in the rejection region, we reject the null hypothesis and accept the research hypothesis.

Two kinds of errors may occur.

1. False rejection of the null hypothesis. The probability of this error is denoted by α.

2. False acceptance of the null hypothesis.

Usually we have to predetermine the value of α. For example, by setting $\alpha = \dfrac{1}{20} = 0.05$, we will be incorrectly rejecting the null hypothesis one time in 20.

EXAMPLE:

The research hypothesis is that the mean yield per $1,000 is greater than $125. Hence, the null hypothesis is $\mu = 125$. A sample of $n = 49$ deposits leads to the average yield $x = \$138$ and $s = 37$. Is the research hypothesis true?

We shall set $\alpha = 0.025$.

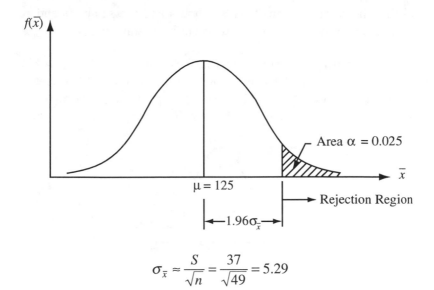

$$\sigma_{\bar{x}} \approx \frac{S}{\sqrt{n}} = \frac{37}{\sqrt{49}} = 5.29$$

For $\alpha = 0.025$, we shall reject the null hypothesis if $\bar{x}$ lies more than 1.96 standard deviations above $\mu = 125$.

$$z = \frac{\bar{x} - \mu}{\sigma_{\bar{x}}} = \frac{138 - 125}{5.29} = 2.46.$$

We see that the observed value of $\bar{x}$ lies more than 1.96 standard deviations above the mean of 125. Hence, we reject the null hypothesis and accept the research hypothesis that the mean yield on a $1,000 investment is greater than $125.

Because the rejection region is located in only one tail (slope), the above test is called a one-tailed test. For the research hypothesis $\mu \neq \mu_0$, we formulate a two-tailed test.

Problem Solving Examples:

Suppose you are a buyer of large supplies of light bulbs. You want to test, at the 5% significance level, the manufacturer's claim that his bulbs last more than 800 hours. You test 36 bulbs and find that the sample mean $\overline{X}$ is 816 hours and the sample standard deviation $s = 70$ hours. Should you accept his claim?

A Establish the hypotheses $H_0:\mu = 800$ hours, and $H_1:\mu > 800$ hours. We know by the Central Limit Theorem that the sampling distribution of the sample means is approximately normal, because $n = 36 > 30$. The rejection area for H_0 is shown in the diagram:

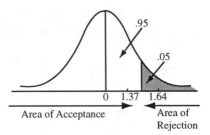

The shaded area represents 5% of the area under the standard normal curve. The table of z-scores gives $z = 1.64$ for a 5% rejection area. Now compute the z-value corresponding to the sample mean $\overline{X} = 816$.

$$z = \frac{\overline{X} - \mu_0}{\sigma_{\overline{x}}},$$

where μ_0 is the mean of the null hypothesis and $\sigma_{\overline{x}}$ is the standard deviation of the sampling distribution of means, which is equal to the quotient of the population standard deviation σ and the square root of the sample size: $\sigma_{\overline{x}} = \sigma/\sqrt{n}$. We do not know the population standard deviation, so we approximate it by the sample standard deviation $s = 70$. Making substitutions in the formula for z, we have

$$z = \frac{\overline{X} - \mu_0}{\sigma_{\overline{x}}} = \frac{\overline{X} - \mu_0}{\dfrac{\sigma}{\sqrt{n}}} = \frac{\overline{X} - \mu_0}{\dfrac{s}{\sqrt{n}}} = \frac{816 - 800}{\dfrac{70}{\sqrt{36}}} = \frac{16}{11.67} = 1.37.$$

This z-value falls in the acceptance region $(-\infty, 1.64)$ for H_0, $\mu = 800$ hours. Therefore, you should reject, at the 5% level, the manufacturer's claim that $\mu > 800$.

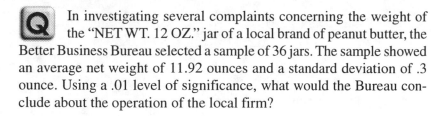

In investigating several complaints concerning the weight of the "NET WT. 12 OZ." jar of a local brand of peanut butter, the Better Business Bureau selected a sample of 36 jars. The sample showed an average net weight of 11.92 ounces and a standard deviation of .3 ounce. Using a .01 level of significance, what would the Bureau conclude about the operation of the local firm?

We use a one-tailed test because we are concerned with whether the actual population mean is at least 12 ounces or whether it is less than 12 ounces. Therefore, we have

$$H_0{:}\mu \geq 12; \; H_1{:}\mu < 12.$$

The figure below depicts this problem. Since the sample size, n, is large (≥ 30), the test statistic $z = (\bar{x} - \mu)/s_{\bar{x}}$ is normally distributed with a mean of 0 and a standard deviation of 1.

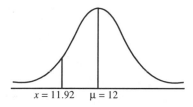

$x = 11.92$ $\mu = 12$

We must calculate

$$z = \frac{\bar{x} - \mu}{s_{\bar{x}}} \; \text{where} \; s_{\bar{x}} = \frac{s}{\sqrt{n}}$$

and compare the obtained value of z to a critical value. Our critical value of z for this problem is -2.33 since 1% of scores of the standard normal distribution have a z-value below -2.33 and we want to reject H_0 for values of $\bar{x}$ less than μ.

Therefore, our decision rule is: reject H_0 if $z < -2.33$; accept H_0 if $z \geq -2.33$.

For the data of this problem

$$s_{\bar{x}} = \frac{.3}{\sqrt{36}} = \frac{.3}{6} = .05$$

and

$$z = \frac{11.92 - 12}{.05} = \frac{-.08}{.05} = -1.60.$$

Since $-1.60 \geq -2.33$, we accept H_0 and conclude that at a 1% level of significance the actual population mean of the local brand of peanut butter is 12 ounces.

 An official of a trade union reports that the mean yearly wage is \$8,000. A random sample of 100 employees in the union produced a mean of \$7,875 with a standard deviation of \$1,000. Test the null hypothesis at the .05 level of significance that the mean wage is \$8,000 against the alternate hypothesis that the wage is greater than or less than \$8,000.

A For this problem H_0 and H_1 are given as follows:

$$H_0: \mu = \$8,000; \ H_1: \mu \neq \$8,000.$$

For $\alpha = .05$, the critical value is $z = 1.96$, and we accept H_0 if z lies between -1.96 and 1.96. This is because the statistic $z = (\overline{X} - \mu)/s_{\overline{X}}$ which has a t-distribution may be approximated by the normal distribution when the sample size, n, is large (≥ 30). For the standard normal distribution, 2.5% of scores will have a z-value above 1.96 and 2.5% of scores will have a z-value below -1.96. We calculate z using the formula:

$$z = \frac{\overline{X} - \mu}{s_{\overline{X}}}$$

where $\overline{X} = 7,875$, $\mu = 8,000$

and

$$s_{\overline{X}} = \frac{s}{\sqrt{n}} = \frac{1,000}{\sqrt{100}} = \frac{1,000}{10} = 100.$$

Therefore, z becomes

$$z = \frac{7,875 - 8,000}{100} = \frac{-125}{100} = -1.25.$$

Since -1.25 lies between -1.96 and 1.96, we accept H_0 that the mean wage is in reality $8,000.

Quiz: Sampling Theory – Statistical Inference

1. An extremely large population has a mean of 16 and a standard deviation of 8. Consider all the possible samples of size 100. What would be the value of the standard deviation of the sampling distribution of the means?

    (A) .80   (D) .16

    (B) .50   (E) .08

    (C) .24

Assume that old tires sold by the Goodmonth Corporation are normally distributed with a mean life of 43,000 miles and a standard deviation of 2,000 miles.

2. If Karen buys six tires, approximately what is the probability that they will average less than 41,500 miles?

    (A) .23   (D) .39

    (B) .27   (E) .43

    (C) .35

3. For which *two* of the following are the statement and the hypotheses correctly matched? (H_0 = null hypothesis, H_1 = alternative hypothesis)

    I.   The mean age of all accountants is at least 34. H_0: $\mu \geq 34$ and H_1: $\mu < 34$.
    II. The proportion of all adults owning credit cards is less than .70. H_0:$p \leq .70$ and H_1:$p > .70$.
    III. The variance for waiting time in a bank line is more than five minutes. H_0: $s^2 \leq 5$ and H_1: $s^2 > 5$.

IV. The mean height, μ_1, for basketball players is greater than the mean height, μ_2, for football players. $H_0: \mu_1 = \mu_2$ and $H_1: \mu_1 \neq \mu_2$.

(A) II and IV only   (D) I and II only

(B) III and IV only   (E) II and III only

(C) I and III only

4. The Alpha car manufacturer claims that their automobiles get at least 10 miles per gallon more than the automobiles from the Beta car manufacturer. Let μ_1 = mean miles per gallon for Alpha cars and μ_2 = mean miles per gallon for Beta cars. If hypothesis testing was conducted for the Alpha car manufacturer's claim, which of the following would be the alternative hypothesis?

(A) $\mu_1 + \mu_2 < 10$ (D) $\mu_1 - \mu_2 < 10$

(B) $\mu_1 - \mu_2 > 10$ (E) $\mu_1 - \mu_2 = 10$

(C) $\mu_1 + \mu_2 > 10$

5. The manager of a department store chain is trying to test whether the mean outstanding balance on 30-day charge accounts is the same in its two suburban branch stores. A random sample of 100 such accounts in each store was taken. In the first store, the mean outstanding balance was $65, and the standard deviation was $15. In the second store, the corresponding numbers were $70 and $18. Which of the following levels of significance would be the *lowest* for which the null hypothesis of equal outstanding balances would have to be rejected?

(A) .05   (D) .02

(B) .04   (E) .01

(C) .03

6. A random sample of 150 high school boys is taken to determine what proportion watches at least an hour of cartoons on television each week. If 40% of the sample watch television, what is the con-

fidence interval at the 95% confidence level for the true percentage of high school boys who watch at least an hour of cartoons?

(A) $0.39 < p < 0.41$

(D) $0.34 < p < 0.46$

(B) $0.30 < p < 0.40$

(E) $0.32 < p < 0.48$

(C) $0.35 < p < 0.45$

7. Two independent samples of size 100 are studied and their means compared. At the 0.04 level of significance, what is the critical statistic, or key value of z, that determines whether or not the difference between the means is significant?

(A) ±1.75

(D) ±2.33

(B) ±1.96

(E) ±0.04

(C) ±2.05

8. In an experiment concerning a mean value with a sample of size $n = 100$, the limits of the confidence interval are 16.6 and 18.0. If the standard deviation is 3.0, at approximately what degree of confidence was the experiment done?

(A) 90%

(B) 95%

(C) 98%

(D) 99%

(E) You cannot tell from the given information.

9. Given an extremely large size for a population, suppose a sample of this population has 144 data, with a mean of 18 and a standard deviation of 6. With what percent confidence (approximately) can it be stated that the actual mean of the population lies between 17.5 and 18.5?

(A) 83

(D) 68

(B) 78

(E) 63

(C) 73

10. Suppose an extremely large population has a mean of 100 and a standard deviation of 8. If the sample means of all samples of size 9 were extracted, what would be the value of the standard deviation of this group of sample means?

(A) $\dfrac{9}{100}$

(D) $\dfrac{100}{9}$

(B) $\dfrac{8}{9}$

(E) $\dfrac{100}{3}$

(C) $\dfrac{8}{3}$

ANSWER KEY

1. (A)   6. (E)

2. (A)   7. (C)

3. (C)   8. (C)

4. (D)   9. (D)

5. (B)   10. (C)

CHAPTER 10

Statistical Inference:
Small Samples

10.1 Small Samples

In Chapter 9, situations were discussed where a random sample of 30 or more observations was possible, but that is not always the case. For example, if you gather data about earthquakes it would be rather difficult to obtain a sample of 30 or more. For large samples with $n \geq 30$, the sampling distributions were approximately normal. For small samples with $n < 30$, this approximation is inaccurate and becomes worse as the number of measurements n decreases.

10.1.1 Small Sampling Theory

A study of sampling distributions of statistics for small samples is called **small sampling theory**. The results hold for small and large samples. Two important distributions used are:

1. Student's t Distribution

2. Chi-Square Distribution.

10.2 Student's *t* Distribution

For small sample sizes, the test statistic

$$z = \frac{\bar{x} - \mu_0}{\sigma} \sqrt{n} \tag{1}$$

with σ replaced by s is falsely rejecting the null hypothesis $\mu = \mu_0$ at a higher percentage than that specified by α.

W.S. Gosset found the distribution and percentage points of the test statistic

$$\frac{\bar{x} - \mu_0}{s} \sqrt{n}, \quad n < 30$$

He published the results under the pen name "Student."

Define the statistic

$$t = \frac{\bar{x} - \mu}{s} \sqrt{n}. \tag{2}$$

Now consider samples of size n drawn from a normal population with mean μ. For each sample, compute t using the sample mean $\bar{x}$ and sample standard deviation s. Thus, we obtain the sampling distribution for t, which is given by

$$Y = \frac{Y_0}{\left(1 + \dfrac{t^2}{n-1}\right)^{\frac{n}{2}}} = \frac{Y_0}{\left(1 + \dfrac{t^2}{v}\right)^{\frac{v+1}{2}}} \tag{3}$$

where Y_0 is a constant depending on the sample size n and such that the area under the curve (3) is equal to 1.

The constant

$$v = n - 1 \tag{4}$$

(sometimes denoted df) is called the number of degrees of freedom.

The distribution (3) is called Student's t distribution.

Note that for large values of n, curve (3) approaches the standardized normal curve.

$$Y = \frac{1}{\sqrt{2\pi}} e^{-\frac{t^2}{2}} \qquad (5)$$

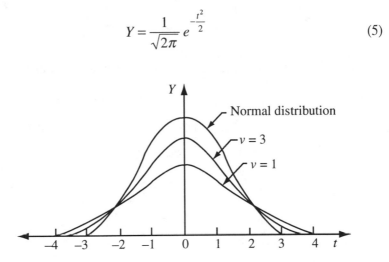

There are many different t distributions. We choose a particular one by specifying the parameter v (the number of degrees of freedom).

As n, and hence v, gets larger, the t distribution approaches the z distribution.

10.2.1 Confidence Intervals

Using the tables of the t distribution, we can define 90%, 95%, 99%, and other confidence intervals, as was done for the normal distributions. The population mean μ can be estimated within specified limits of confidence.

Let us denote by $-t_{0.95}$ and $t_{0.95}$ the values of t for which 5% of the area lies in each tail of the t distribution. Then, a 90% confidence interval for t is

$$-t_{0.95} < \frac{\bar{x} - \mu}{s}\sqrt{n} < t_{0.95}. \tag{6}$$

From (6) we find

$$\bar{x} - t_{0.95} < \frac{s}{\sqrt{n}} < \mu < \bar{x} + t_{.95}\frac{s}{\sqrt{n}}. \tag{7}$$

Hence, μ lies in the interval $\bar{x} \pm t_{0.95}\dfrac{s}{\sqrt{n}}$ with 90% confidence.

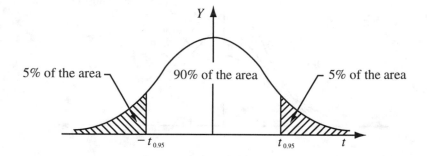

In general, the confidence limits for population means are defined by

$$\bar{x} \pm t_c \frac{s}{\sqrt{n}} \tag{8}$$

where the values $\pm t_c$ are called critical values or confidence coefficients.

We choose t_c depending on the confidence desired and the sample size.

EXAMPLE:

The graph of Student's t distribution with 7 degrees of freedom $(\nu = 7)$ is shown

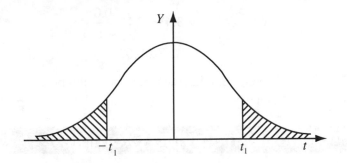

We want to find the value of t_1, for which the total shaded area is 0.05. Student's t distribution is symmetrical. Thus, the shaded area to the right of t_1 is 0.025. The area to the left is

$$1 - 0.025 = 0.975.$$

t_1 represents the 97.5th percentile $t_{0.975}$.

From the tables for $v = 7$, we find

$$t_{0.975} = 2.36.$$

The required value of t is 2.36.

EXAMPLE:

A sample of 17 measurements of the diameter of a drill gave a mean $\bar{x} = 3.47$ inches and a standard deviation $s = 0.05$ inches.

We shall find 95% confidence limits for the actual diameter of the drill. The 95% confidence limits are given by

$$\bar{x} \pm t_{0.975} \frac{s}{\sqrt{n}}.$$

Since $n = 17$, the number of degrees of freedom v is

$$v = n - 1 = 16.$$

From the tables for $v = 16$ and $t_{0.975}$, we find

$$t_{0.975} = 2.12.$$

Substituting all values, we find

$$\bar{x} \pm t_{0.975} \frac{s}{\sqrt{n}} = 3.47 \pm 2.12 \frac{0.05}{4.123}$$
$$= 3.47 \pm 0.0257.$$

Hence, we are 95% confident that the actual mean lies between 3.4443 inches and 3.4957 inches.

Problem Solving Examples:

 A sample of 10 patients in the intensive care unit of Brothers' Hospital in Frostbite Falls, Minnesota has a mean body temperature of 103.8°. The standard deviation is 1.26 degrees. Establish a 95% confidence interval for μ, the mean temperature of Brothers' intensive care patients.

We have a small sample and an unknown variance. The t distribution is called for. We know

$\dfrac{\bar{X} - \mu}{s/\sqrt{n}}$ is t distributed with $n - 1$ degrees of freedom. Here we are

told $\bar{X} = 103.8$, $s = 1.26$, and $n = 10$. Hence, $n - 1 = 9$ and $\sqrt{n} = 3.16$. From the t tables we can see that $P(-2.262 < t_{(9)} < 2.262) = .95$.

With our $t_{(9)}$, the inequality within the parentheses is -2.262

$$< \frac{\bar{X} - \mu}{s/\sqrt{n}} < 2.262.$$

Substituting our given values:

$$-2.262 < \frac{103.8 - \mu}{1.26/3.16} < 2.262.$$

Multiplying through by $\dfrac{1.26}{3.16}$, $-.902 < 103.8 - \mu < .902$.

Subtracting 103.8, $-104.702 < -\mu < -102.898$.

Multiply by -1, $102.898 < \mu < 104.702$.

The mean temperature, with 95% confidence, is between 102.898 and 104.702.

Using the information in the previous problem, establish a 99% confidence interval.

 The value of 3.25 is pulled from a *t*-distribution table under the column for .005 (two tails) and for 9 degrees of freedom. (Even though $n = 10$, the number of degrees of freedom is always $n - 1$).

$$-3.25 < \frac{\overline{X} - \mu}{s/\sqrt{n}} < 3.25$$

Then $\qquad -3.25 < \dfrac{103.8 - \mu}{1.26/3.16} < 3.25$

Simplifying, $\quad -1.296 < 103.8 - \mu < 1.296$

Finally, $\qquad\quad 102.504 < \mu < 105.096.$

10.3 Tests of Hypotheses and Significance

The tests of hypotheses and significance for small samples are similar to those for large samples. The *z* score of *z* statistic is replaced for small samples with the appropriate *t* score or *t* statistic.

10.3.1 Small-Sample Test for m

Research hypothesis *h*:

1. $\mu > \mu_0$

2. $\mu < \mu_0$

3. $\mu \neq \mu_0$

Corresponding Null hypothesis h_0 :

1. $\mu \leq \mu_0$

2. $\mu \geq \mu_-$

3. $\mu = \mu_0$

Test statistic

$$t = \frac{\bar{x} - \mu}{s} \sqrt{n}$$

where $\bar{x}$ is the mean of the sample of size n.

Rejection region: For a probability α of a type 1 error and $\nu = n - 1$,

    1. reject h_0 if $t > t_{\alpha}$

    2. reject h_0 if $t < -t_{\alpha}$

    3. reject h_0 if $|t| > t_{\frac{\alpha}{2}}$

EXAMPLE:

A manufacturer claims that the mean lifetime of his tires is 75,000 miles. A test was performed on five sets of tires; their mean lifetime was 73,500 miles and a standard deviation $s = 1,250$ miles. We want to verify the manufacturer's claim at a level of significance of 0.05.

The null hypothesis h_0 is $\mu_0 \geq 75,000$. We have

$$t = \frac{\bar{x} - \mu}{s} \sqrt{n} = \frac{73,500 - 75,000}{1,250} \sqrt{5}$$

$$= -2.68.$$

For a one-tailed test at a 0.05 level of significance, we shall decide according to the rules:

1. Accept h_0 if t is greater than $-t_{0.95}$. For $\nu = 5 - 1 = 4$ the value of $-t_{0.95}$ is -2.13.

2. Reject h_0 otherwise. Since $t = -2.68$, we reject h_0.

10.3.2 Differences of Means

Suppose two random samples of sizes n_1 and n_2 are drawn from normal populations with equal standard deviations $\sigma_1 = \sigma_2$. The samples have means x_1, x_2 and standard deviations s_1, s_2, respectively. We formulate the null hypothesis h_0 as follows:

The samples come from the same population, that is,

$$\mu_1 \quad \mu_2 \text{ and } \sigma_1 = \sigma_2 .$$

The t score used is given by

$$t = \frac{\bar{x}_1 - \bar{x}_2}{s\sqrt{\dfrac{1}{n_1} + \dfrac{1}{n_2}}}$$

where

$$s = \sqrt{\frac{(n_1 - 1)s_1^2 + (n_2 - 1)s_2^2}{n_1 + n_2 - 2}}$$

The distribution is Student's t distribution with

$$\nu = n_1 + n_2 - 2$$

degrees of freedom.

EXAMPLE:

The intelligence quotients of two groups of students from two universities were tested. The test should determine if there is a significant difference between the I.Q.'s of the two groups.

The following results were obtained:

For Group A

sample size $n_A = 14$, mean 146 with a standard deviation of 12

For Group B

sample size $n_B = 17$, mean 141 with a standard deviation of 10

Let μ_A and μ_B denote population mean I.Q.'s of students from two universities.

We set the level of significance at 0.05. We have to decide between the two hypotheses:

1. h_0: $\mu_A = \mu_B$, there is no significant difference between the groups.

2. h_1: $\mu_A \neq \mu_B$, there is a significant difference between the groups.

We have

$$\sigma = \sqrt{\frac{(n_A - 1)s_A^2 + (n_B - 1)s_B^2}{n_1 + n_2 - 2}} = \sqrt{\frac{13 \times (12)^2 + 16 \times (10)^2}{14 + 17 - 2}}$$

$$= 10.94.$$

$$t = \frac{\overline{x}_A - \overline{x}_B}{\sigma \sqrt{\frac{1}{n_A} + \frac{1}{n_B}}} = \frac{146 - 141}{10.94 \sqrt{\frac{1}{14} + \frac{1}{17}}}$$

$$= 1.27.$$

For the two-tailed test at a level of significance, we would reject h_0 if t were outside the range $-t_{0.975}$ to $t_{0.975}$. For $v = 29$ this range is -2.04 to 2.04. Hence, we cannot reject h_0 at a level of significance 0.05.

The conclusion is that there is no significant difference between the I.Q.'s of the two groups.

Problem Solving Examples:

 A lathe is adjusted so that the mean of a certain dimension of the parts is 20 cm. A random sample of 10 of the parts produced a mean of 20.3 cm and a standard deviation of .2 cm. Do the results indicate that the machine is out of adjustment? Test at the .05 level of significance.

A Since the question asks whether the machine is out of adjustment but does not ask about the direction of this possible incorrect adjustment, a two-tailed test is appropriate. We use a t-test with a null hypothesis that the mean of the population is 20 cm and an alternate hypothesis that the mean is not equal to 20 cm. Thus, we have

$$H_0: \mu = 20; \ H_1: \mu \neq 20.$$

The data for this problem are illustrated in the following figure.

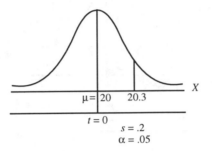

For $n = 10$, we have $df = n - 1 = 9$ degrees of freedom. For 9 df, at a .05 level of significance, and a two-tailed test, the critical value of t is 2.262. This is because of the fact that for a t distribution with a mean of 0 and a standard deviation of 1, 2.5% of scores will have a t value above 2.262 and 2.5% of scores will have a t value below -2.262.

Hence, our decision rule is: Reject H_0 if $t > 2.262$ or $t < -2.262$; accept H_0 if $-2.262 \leq t \leq 2.262$. The test statistic here is

$$t = \frac{\overline{X} - \mu}{\left(s/\sqrt{n}\right)} = \frac{\overline{X} - \mu}{s}\sqrt{n}.$$

For the given values,

$$t = \frac{20.3 - 20.0}{.2}\sqrt{10} = \frac{.3}{.0632} = 4.74.$$

Since $4.74 > 2.262$, we reject H_0 and conclude that the machine is in need of an adjustment.

Q From appropriately selected samples, two sets of IQ scores are obtained. For group 1, $\overline{x} = 104, s = 10$, and $n = 16$; for group 2, $x = 112, S = 8$, and $n = 14$. At the 5% significance level, is there a significant difference between the two groups?

For this problem, we have for our hypotheses $H_0: \mu_1 - \mu_2 = 0$ and $H_1 : \mu_1 - \mu_2 \neq 0$.

This problem can be depicted by the diagram that follows.

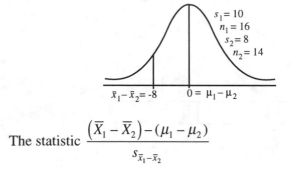

$s_1 = 10$
$n_1 = 16$
$s_2 = 8$
$n_2 = 14$

$\bar{x}_1 - \bar{x}_2 = -8 \qquad 0 = \mu_1 - \mu_2$

The statistic $\dfrac{\left(\overline{X}_1 - \overline{X}_2\right) - \left(\mu_1 - \mu_2\right)}{s_{\bar{x}_1 - \bar{x}_2}}$

has a t distribution when $n_1 + n_2 \leq 30$.

We have $n_1 + n_2 - 2 = 28$ *df*'s and so our decision rule for $\alpha = .05$ is, reject H_0 if $t > 2.048$ or $t < -2.048$; accept H_0 if $-2.048 \leq t \leq 2.048$. We must calculate

$$t = \frac{\left(\overline{X}_1 - \overline{X}_2\right) - \left(\mu_1 - \mu_2\right)}{s_{\bar{x}_1 - \bar{x}_2}}$$

where

$$s_{\bar{x}_1 - \bar{x}_2} = \sqrt{\frac{\left(n_1 - 1\right)s_1^2 + \left(n_2 - 1\right)s_2^2}{n_1 + n_2 - 2}} \sqrt{\frac{1}{n_1} + \frac{1}{n_2}}.$$

For the data of this problem,

$$s_{\bar{x}_1 - \bar{x}_2} = \sqrt{\frac{15(10) + (13)(8)^2}{28}} \sqrt{\frac{1}{16} + \frac{1}{14}}$$

$$= \sqrt{\frac{15(100) + (13)(64)}{28}} \sqrt{.0625 + .0714}$$

$$= \sqrt{\frac{1,500 + 832}{28}} \sqrt{.1339}$$

$$= \sqrt{83.29} \sqrt{.1339} = 3.34$$

and

$$t = \frac{(104 - 112) - 0}{3.34} = \frac{-8}{3.34} = 2.40.$$

Since $-2.40 < -2.048$, we reject H_0 and conclude that there is a significant difference between the scores of the two groups at the 5% level of significance.

10.4 The Chi-Square Distribution

We define the statistic

$$\chi^2 = \frac{(x_1 - \bar{x})^2 + (x_2 - \bar{x})^2 + \ldots + (x_n - \bar{x})^2}{\sigma^2}$$

χ is the Greek letter chi.

Samples of size n are drawn from a normal population with standard deviation σ. For each sample, the value χ^2 is computed. We obtain the chi-square distribution

$$Y = Y_0 \, \chi^{\nu-2} \, e^{-\frac{1}{2}\chi^2}$$

where $\nu = n - 1$ is the number of degrees of freedom, and Y_0 is a constant depending on ν and such that the area under the curve is 1.

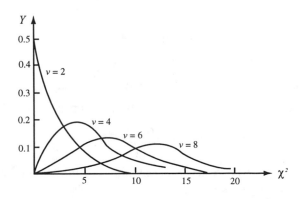

Use the statistic

$$\chi^2 = \frac{(n-1)s^2}{\sigma^2}$$

to test a hypothesis about a population variance σ^2 based on a sample variance s^2. We assume the population is normally distributed.

Problem Solving Examples:

 A sample of size 10 produced a variance of 14. Is this sufficient to reject the null hypothesis that $\sigma^2 = 6$ when tested using a .05 level of significance? Using a .01 level of significance?

 We use the χ^2 (chi-square) statistic to determine the value of a population variance when given a sample variance. We may do this because the test statistic $\frac{(n-1)s^2}{\sigma^2}$ has a c^2 distribution with $n-1$ degrees of freedom. Here we have as our hypotheses:

$$H_0: \sigma^2 = 6, \ H_1: \sigma^2 \neq 6.$$

Since $\alpha = .05$ and this is a two-tailed test, we will reject H_0 if our calculated value for $\chi^2 > \chi^2_{.025}$ or $< \chi^2_{.975}$. We will accept H_0 if our calculated value for χ^2 lies between $\chi^2_{.975}$ and $\chi^2_{.025}$. The number of degrees of freedom for $\chi^2_{.025}$ and $\chi^2_{.975}$ is $n-1$, or, in this case, 9. Therefore, our decision rule is: reject H_0 if calculated $\chi^2 > 19.023$ or $\chi^2 < 2.700$; accept H_0 if $2.700 \leq \chi^2 \leq 19.023$.

We now calculate χ^2 using the formula

$$\chi^2 = \frac{(n-1)s^2}{\sigma^2}.$$

Since $n = 10$, $s^2 = 14$, and $\sigma^2 = 6$ for this problem, we have

$$\chi^2 = \frac{9(14)}{6} = 21.$$

Since $21 > 19.023$, we reject H_0 and conclude that $\sigma^2 \neq 6$ at the 5% level of significance.

For $\alpha = .01$, we must compare our calculated χ^2 to $\chi^2_{.005}$ and $\chi^2_{.995}$, again for 9 degrees of freedom. Our decision rule becomes: reject H_0 if calculated $\chi^2 > 23.589$ or $\chi^2 < 1.735$; accept H_0 if $1.735 \leq \chi^2 \leq 23.589$. Since our calculated χ^2 was 21, and $1.735 \leq 21 \leq 23.589$, we would accept H_0 that $\sigma^2 = 6$ at a 1% level of significance.

Q The makers of a certain brand of car mufflers claim that the life of the mufflers has a variance of .8 year. A random sample of 16 of these mufflers showed a variance of one year. Using a 5% level of significance, test whether the variance of all the mufflers of this manufacturer exceeds .8 year.

A Our hypotheses for this problem are:

$$H_0: \sigma^2 \leq .8, H_1: \sigma^2 > .8.$$

The statistic $\dfrac{(n-1)s^2}{\sigma^2}$ has a χ^2 distribution with $n-1$ degrees of freedom.

For $\alpha = .05$, a one-tailed test, and $n-1 = 15$ degrees of freedom, we will have for our decision rule: reject H_0 if $\chi^2 > 24.996$; accept H_0 if $\chi^2 \leq 24.996$.

We calculate χ^2.

$$\chi^2 = \frac{(n-1)s^2}{\sigma^2} = \frac{15(1)^2}{.8} = 18.75$$

Since $18.75 < 24.996$, we accept H_0 and conclude that the variance of all the mufflers of this manufacturer does not exceed .8 year.

Q With individual waiting lines at its various windows, the manager of a bank finds that the standard deviation for waiting times is 6.2 minutes. An experiment with one waiting line shows that for a random sample of 25 customers, the waiting times have a standard deviation of 4.2 minutes. At the .05 level of significance, test the hypothesis that a single line causes lower variation among waiting times.

We have $n = 25$, $s = 4.2$.

H_0: $\sigma \geq 6.2$ is the null hypothesis.

H_A: $\sigma < 6.2$ is the alternative hypothesis.

Now $\chi^2 = \dfrac{(25-1)(4.2)^2}{(6.2)^2} \approx 11.01$ is the test statistic. The critical χ^2 value for 24 degrees of freedom with an area of .05 to its left is 13.848. Since $11.01 < 13.848$, the test statistic falls within the critical region. Thus, the sample data support the alternative hypothesis claim of lower variation.

A software firm claims that the times required to run its computer programs have a standard deviation of 25 hours. A sample of 30 computer programs showed a standard deviation of 28 hours. At the .05 level of significance, test the firm's claim.

We have $n = 30$, $s = 28$.

H_0: $\sigma = 25$ is the null hypothesis

H_A: $\sigma \neq 25$ is the alternative hypothesis.

Now $\chi^2 = \dfrac{(30-1)(28)^2}{(25)^2} \approx 36.38$ is the test statistic. The critical χ^2 values for 29 degrees of freedom with an area of .025 to its left and .025 to its right are 16.047 and 45.722, respectively. Since 36.38 falls between 16.047 and 45.722, we cannot reject the null hypothesis that $\sigma = 25$.

10.5 Control Charts

From the manufacturer's and consumer's point of view, it is important to maintain a steady level of quality of a product.

Control charts are used to monitor the quality of a product and to detect possible shifts in quality. For samples collected over a period of time, we graph the sample mean or sample range.

A control chart consists of three lines: a center line, an upper line, and a lower line. The means of successive samples are plotted on the chart.

The center line, denoted by $\bar{x}_c$, represents the average of k sample means each computed from m observations. Generally, we take

$$k \geq 25 \quad \text{and}$$

$$n \geq 4.$$

The samples are taken at the time when production is judged to be normal.

By x_{ij} we denote the jth observation in sample i.

$$\bar{x}_c = \frac{\sum_i \sum_j x_{ij}}{km}$$

The upper control line (UCL) is computed from

$$\text{UCL} = \bar{x}_c - 3\frac{\sigma}{\sqrt{n}}$$

and the lower control line (LCL)

$$\text{LCL} = \bar{x}_c - 3\frac{\sigma}{\sqrt{n}}$$

The interval $\bar{x}_c \pm 3\frac{\sigma}{\sqrt{n}}$ should contain almost all the sample means $\sum_j x_{ij}$ in repeated sampling.

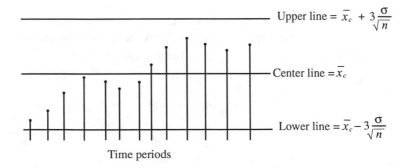

Upper line $= \bar{x}_c + 3\frac{\sigma}{\sqrt{n}}$

Center line $= \bar{x}_c$

Lower line $= \bar{x}_c - 3\frac{\sigma}{\sqrt{n}}$

Time periods

CHAPTER 11

Mathematical Models
Relating Independent and
Dependent Variables

11.1 Relationships Between Variables

Often in practice, a relationship exists between two or more variables. The dependent variable (the variable of interest) can be affected by one or more independent variables. We usually try to express this relationship in the mathematical form of an equation. The simplest one is the linear equation (equation of a straight line).

Straight Line

$$y = a_0 + a_1 x$$

The response y is related to a single quantitative independent variable x. Here, a_0 and a_1 are constants. This model is a deterministic model; the value of y is uniquely determined by the value of x.

Parabola

$$y = a_0 + a_1 x + a_2 x^2$$

Relationships Between Variables

Cubic Curve

$$y = a_0 + a_1 x + a_2 x^2 + a_3 x^3$$

*n*th Degree Curve

$$y = a_0 + a_1 x + \ldots + a_n x^n$$

In all these cases, a response y is related to a single independent variable. Note that the right-hand side of all equations are polynomials.

There are many other possible relationships between x and y.

Hyperbola	$y = \dfrac{1}{a_0 + a_1 x}$
Exponential curve	$y = ab^x$
Geometric curve or power curve	$y = ax^b$
Logistic curve	$y = \dfrac{1}{ab^x + d}$

In practice, it often happens that the relationship between x and y cannot be expressed in the form of an equation. The data can be presented in a scatter diagram.

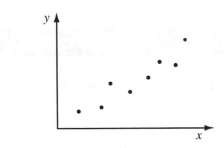

Eight points are shown in the diagram; their distribution resembles a straight line. We introduce the following model

$$y = a_0 + a_1 x + \varepsilon$$

where ε is a random error; here, it represents the difference between a measurement y and a point on the line $a_0 + a_1 x$. This model is called the probabilistic model.

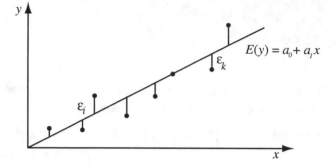

The average of y, called the expected value of y for a fixed x, is

$$E(y) = a_0 + a_1 x$$

If two points

$$(x_1, y_1) \text{ and } (x_2, y_2)$$

on the line are given, then the constants in $a_0 + a_1 x = y$ can be determined. The equation of the line can be written

$$y - y_1 = \frac{(y_2 - y_1)}{x_2 - x_1}(x - x_1)$$

Problem Solving Examples:

 Write the equation of a line containing (1, 3) and (4, 9).

 $y - 3 = \left(\dfrac{9-3}{4-1}\right)(x-1)$. This will simplify to $y = 2x + 1$.

 Write the equation of an exponential curve in the form of $y = ab^x$ which contains the points (1, 2) and (3, 50).

 $2 = ab^1$ and $50 = ab^3$. Dividing the second equation by the first equation, $25 = b^2$, so $b = 5$ or -5. If $b = 5$, $a = {}^2/_5$. If $b = -5$, $a = {}^{-2}/_5$. The two possible equations are $y = ({}^2/_5)(5^x)$ and $y = ({}^{-2}/_5)(-5)^x$.

11.2 Probabilistic Models

The simplest type of probabilistic model is

$$y = a_0 + a_1 x + \varepsilon.$$

The assumption is that the average value of ε for a given value of x is

$$E(\varepsilon) = 0.$$

Hence, the expected value of y for a given value of x is

$$E(y) = a_0 + a_1 x.$$

Not all data would fit this model. There are situations that require

$$y = a_0 + a_1 x + a_2 x^2 + \varepsilon$$

with

$$E(y) = a_0 + a_1 x + a_2 x^2.$$

A general polynomial probabilistic model relating a single independent variable x to a dependent variable y is given by

$$y = a_0 + a_1 x + a_2 x^2 + \ldots + a_n x^n + \varepsilon$$

with

$$E(y) = a_0 + a_1 x + \ldots + a_n x^n$$

The choice of n depends on the experimental situation.

11.2.1 Models with Several Independent Variables

Sometimes, y depends on more than one independent variable. Here are some typical probabilistic models relating a response y to two independent variables, x_1 and x_2:

$$y = a_0 + a_1 x_1 + a_2 x_2 + \varepsilon$$

$$y = a_0 + a_1 x_1 + a_2 x_2 + a_3 x_1 x_2 + \varepsilon$$

$$y = a_0 + a_1 x_1 + a_2 x_1^2 + a_3 x_2 + \varepsilon$$

Similarly, we can construct models relating y to three or more independent variables.

The general linear model is

$$y = a_0 + a_1 x_1 + a_2 x_2 + \ldots + a_n x_n + \varepsilon$$

where $x_1, \ldots, x_n$ are independent variables, $a_0, a_1, \ldots, a_n$ are unknown parameters, and ε is a random error with

$$E(\varepsilon) = 0.$$

All probabilistic models discussed here are special cases of the general linear model. For example, the model

$$y = a_0 + a_1 x + a_2 x^2 + a_3 x^3 + \varepsilon$$

is equivalent to a linear model

$$y = a_0 + a_1 x_1 + a_2 x_2 + a_3 x_3 + \varepsilon$$

where $x = x_1$, $x^2 = x_2$, $x^3 = x_3$.

Individual terms in the general model are classified by their exponents. The degree of a term is equal to the sum of the exponents for the variables in this term. Thus,

$$6x_4{}^7 \text{ is a 7th-degree term}$$

and

$$x_3{}^4 x_4{}^5 \text{ is a 9th-degree term.}$$

A first-order model is a general linear model that contains all possible first-degree terms in the independent variables.

11.3 The Method of Least Squares

Let $x_1, x_2, \ldots, x_n$ be independent variables and y a response:

$$y = a_0 + a_1 x_1 + a_2 x_2 + \ldots + a_n x_n + \varepsilon.$$

We assume that the random error has expectation zero and obtain the expected value of y

$$E(y) = a_0 + a_1 x_1 + \ldots + a_n x_n.$$

This line is called the regression of y on $x_1, \ldots, x_n$. In real situations, the parameters $a_0, a_1, \ldots, a_n$ are not known. We cannot find $E(y)$, but we can construct an estimate of $E(y)$ by using the equation

$$\hat{y} = \hat{a_0} + \hat{a_1} x_1 + \ldots + \hat{a_n} x_n$$

where $\hat{a_0}, \hat{a_1}, \ldots + \hat{a_n}$ are estimates of the unknown parameters $a_0, a_1, \ldots, a_n$. We find these estimates from sample data.

Consider the probabilistic model

$$y = a_0 + a_1 x + \varepsilon$$

for the linear regression

$$E(y) = a_0 + a_1 x.$$

There are many ways of finding an estimate of $E(y)$

$$\hat{y} = \hat{a_0} + \hat{a_1} x.$$

One can draw an approximating curve to fit a set of data. This method is called a freehand method of curve fitting.

The more reliable method is the method of least squares.

Let $|\hat{y}|$ denote the predicted value of y for a given value of x.

We define

$$\text{residual} = y - \hat{y}.$$

The method of least squares finds the prediction line

$$\hat{y} = \hat{a_0} + \hat{a_1} x$$

that minimizes the value of

$$\sum \left(y - \hat{y} \right)^2.$$

The sum is taken over all sample points. For the linear model

$$y = a_0 + a_1 x + \varepsilon$$

this sum is equal to

$$\sum \left(y - \hat{y} \right)^2 = \sum \left(y - \hat{a}_0 - \hat{a}_1 x \right)^2.$$

The method of least squares is based on finding the estimates of $\hat{a}_0$ and $\hat{a}_1$ which minimize

$$\sum \left(y - \hat{y} \right)^2.$$

The results are summarized below.

Least squares estimates of a_0 and a_1

$$\hat{a}_1 = \frac{\sum (x - \bar{x})(y - \bar{y})}{\sum (x - \bar{x})^2}$$

$$\hat{a}_1 = \bar{y} - \hat{a}_1 \bar{x}$$

Note that

$$\sum (x - \bar{x})_2 (y - \bar{y}) = \sum xy - \frac{\left(\sum x \right)\left(\sum y \right)}{n}$$

$$\sum (x - \bar{x})^2 = \sum x^2 - \frac{\left(\sum x \right)^2}{n}$$

EXAMPLE:

In some statistical experiments, a random sample of size $n = 9$ was chosen and the corresponding values of x and y were measured.

x	y	x^2	xy
37	31	1,369	1,147
25	19	625	475
32	26	1,024	832
18	12	324	216
29	27	841	783
34	28	1,156	952
23	19	529	437
28	21	784	588
36	30	1,296	1,080

$$\sum x = 262 \quad \sum y = 213 \quad \sum x^2 = 7{,}948 \quad \sum xy = 6{,}510$$

We shall construct a straight line which approximates the data of the table. We are looking for the equation of the line

$$y = a_0 + a_1 x$$

We have

$$a_1 = \frac{\sum (x - \bar{x})(y - \bar{y})}{\sum (x - \bar{x})^2} = \frac{n \sum xy - (\sum x)(\sum y)}{n \sum x^2 - (\sum x)^2}$$

$$= \frac{9 \times (6510) - (262) \times (213)}{9 \times (7948) - (262)^2}$$

$$= \frac{2784}{2888} = 0.964$$

$$a_0 = \bar{y} - a_1 \bar{x} = \frac{(\sum x^2)(\sum y^2) - (\sum x)(\sum xy)}{n \sum x^2 - (\sum x)^2}$$

$$= \frac{213}{9} - (.964)\left(\frac{262}{9}\right)$$

$$= -4.396$$

Thus,

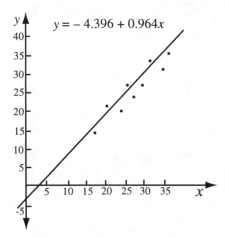

$$y = -4.396 + 0.964x$$

Plot of the Least Squares Equation

Consider again the equation

$$y = a_0 + a_1 x$$

from which we can obtain the normal equations

$$\left.\begin{array}{l} \sum y = na_o + a_1 \sum x \\ \sum xy = a_0 \sum x + a_1 \sum x^2 \end{array}\right\}$$

Taking the data from the table, we find

$$\left.\begin{array}{l} 213 = 9a_0 \quad + 262a_1 \\ 6510 = 262a_0 + 7948a_1 \end{array}\right\}$$

Solving this system simultaneously, we find

$$a_0 = -4.396$$

$$a_1 = 0.964$$

11.3.1 The Least Square Parabola

Suppose the measurements are

$$(x_1, y_1)\ (x_2, y_2)\ ...\ (x_n, y_n)$$

where x is an independent variable. The least square parabola has the equation

$$E(y) = a_0 + a_1 x + a_2 x^2$$

from which we find

$$\left.\begin{aligned}
\sum y &= a_0 n + a_1 \sum x + a_2 \sum x^2 \\
\sum xy &= a_0 \sum x + a_1 \sum x^2 + a_2 \sum x^3 \\
\sum x^2 y &= a_0 \sum x^2 + a_1 \sum x^3 + a_2 \sum x^4
\end{aligned}\right\}$$

This system is called the system of the normal equations for the least square parabola.

Solving simultaneously the normal equations, we find $a_0,\ a_1,\ a_2$.

11.3.2 General Linear Models

We can apply the method of least squares to estimate parameters of the general linear model. Let

$$y = a_0 + a_1 x_1 + a_2 x_2 + ... + a_k x_k + \varepsilon.$$

We look for estimates $\hat{a}_0, \hat{a}_1, ..., \hat{a}_k$ of $a_0, a_1, ..., a_k$, respectively, which minimize the value of

$$\sum (y - \hat{y})^2 = \sum (y - \hat{a}_0 - \hat{a}_1 x_1 - \hat{a}_2 x_2 - ... - \hat{a}_k x_k)^2.$$

It can be shown that

$$\sum (y - \bar{y})^2 = \sum (y - y)^2 + \sum (y - \bar{y})^2.$$

Problem Solving Example:

 Given four pairs of observations

X	6	7	4	3
Y	8	10	4	2

compute and graph the least-squares regression line.

 We use the table to compute the summary statistics needed.

X	Y	X^2	XY	
6	8	36	48	
7	10	49	70	
4	4	16	16	
3	2	9	6	
$\sum X = 20$	$\sum Y = 24$	$\sum X^2 = 110$	$\sum XY = 140$	$n = 4$

From the normal equations $\sum y = na_0 + a_1 \sum x$, $\sum xy = a_0 \sum x + a_1 \sum x^2$, we find $24 = 4a_0 + 20a_1$, and $140 = 20a_0 + 110a_1$, solving simultaneously, $a_0 = -4$ and $a_1 = 2$. So, $\hat{y} = -4 + 2x$.

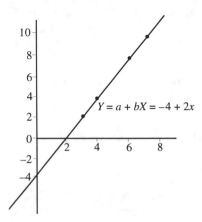

CHAPTER 12

General Linear Model Inferences: Correlation Theory

12.1 Introduction

A set consisting of experimental data is given. The problem (situation) is described by the general linear model with a prescribed number of independent variables. Using the data, we find least square estimates of parameters in this model. The next step is to make inferences about the corresponding population parameters.

12.2 Single Parameter

Consider the general linear model with k independent variables and n experimental data

$$y_i = a_0 + a_1 x_{i1} + \ldots + a_k x_{ik} + \varepsilon_i$$

$$i = 1, 2, \ldots, n$$

We assume that

1. The random error ε_i of observation i has expectation 0.

2. Random errors $\varepsilon_1, \ldots, \varepsilon_n$ are independent of each other.

3. $\varepsilon = [\varepsilon_1, \ldots, \varepsilon_n]$ has mean 0 and variance σ^2.

From the sample data, we obtain an estimate of σ^2.

For the general linear model

$$y = a_0 + a_1 x_1 + \ldots + a_k x_k + \varepsilon_i, \; E(\varepsilon) = 0$$

we have

$$\varepsilon = y - E(y).$$

Since $E(y)$ is not known, we can estimate $E(y)$ from $\hat{y}$.

The residual sum of squares (also called the sum of squares for error) is defined as

$$\sum (y - \hat{y})^2.$$

Dividing $\sum (y - \hat{y})^2$ by the number of degrees of freedom, we find an estimate of σ^2.

The number of degrees of freedom is equal to the sample size n minus the number of parameters $k + 1$. Hence,

$$\text{Number of degrees of freedom} = n - (k + 1).$$

We have an estimate of σ^2 from the general linear model

$$s^2 = \frac{\sum (y - \hat{y})^2}{n - (k + 1)}.$$

Usually, we assume that ε_i's in the general linear model are normally distributed. Hence, for a specific parameter we can determine

$$100 \, (1 - \alpha)\%$$

confidence interval

$$\hat{a}_i \pm t_{\frac{\alpha}{2}} Sa_i, \text{ where } S^2 a_i = \frac{Se^2}{\sqrt{\sum_{i=1}^{n}(x_{ii} - \bar{x}_i)^2 (1 - r^2)}}$$

and t is calculated for $n - (k + 1)$ degrees of freedom.

$$s^2 = \frac{\sum (y - \hat{y})^2}{n - (k+1)}$$

We can construct a test of a hypothesis concerning a general linear model parameter a_i. The null hypothesis is $a_i = 0$. There are three research hypotheses $(a_i > 0, a_i < 0, a_i \neq 0)$. We choose one depending on a particular experimental situation.

Test of an Hypothesis Concerning a_i

Research Hypothesis: 1. $a_i > 0$

2. $a_i < 0$

3. $a_i \neq 0$

Null Hypothesis h_0: $a_i = 0$

Test Statistic: $t = \dfrac{\hat{a}_i}{Sa_i}$

Rejection Region: For a given value of α and the number of degrees of freedom, which equals $n - k - 1$,

1. if $t > t_\alpha$, reject h_0

2. if $t < -t_\alpha$, reject h_0

3. if $t > \left| t_{\frac{\alpha}{2}} \right|$, reject h_0

12.2.1 Estimates of E(y)

For a specific setting of

$$x_1, x_2, \ldots, x_k$$

we shall find the estimate of $E(y)$. Consider the prediction equation of that setting

$$\hat{y} = \hat{a}_0 + \hat{a}_1 x_1 + \ldots + \hat{a}_k x_k.$$

Taking the repeated sampling at this specific setting, we obtain the sampling distribution of $\hat{y}$ which has a mean

$$E(y) = a_0 + a_1 x_1 + \ldots + a_n x_n.$$

Assuming that the distribution of the random errors ε_i is normal, we find

$$100 \, (1 - \alpha)\%$$

confidence interval for $E(y)$.

The value of t is calculated for $n - (k + 1)$ degrees of freedom.

To find the $100(1 - \alpha)\%$ confidence interval for $E(y)$, let

$$\hat{y} = a_0 + a_1 x$$

be the prediction equation. For a specific x_0 value, the prediction interval for the actual y value with $100(1 - \alpha)\%$ confidence is given by

$$\hat{y} - E < y < \hat{y} + E$$

where

$$E = t_{\alpha/2} S_y \sqrt{1 + \frac{1}{n} + \frac{n(x_0 - \bar{x})^2}{n\left(\sum x^2\right) - \left(\sum x\right)^2}}$$

Note that $t_{\alpha/2}$ has $n-2$ degrees of freedom. Also, an easier computational formula for s_y is given by $\sqrt{\dfrac{\sum y^2 - a_0 \sum y - a_1 \sum xy}{n-2}}$

For a given setting of the independent variables, we can design a statistical test of $E(y)$.

Research Hypothesis:
1. $E(y) > \mu_0$

2. $E(y) < \mu_0$

3. $E(y) \neq \mu_0$

Null Hypothesis h_0:
$E(y) = \mu_0$

Test Statistic:
$$t = \frac{\hat{y} - \mu_0}{s_y \sqrt{1 + \dfrac{1}{n} + \dfrac{n(x_0 - \bar{x})^2}{n\left(\sum x^2\right) - \left(\sum x\right)^2}}}$$

Rejection Region:
1. if $t > t_\alpha$, reject h_0

2. if $t < t_\alpha$, reject h_0

3. if $t > \left| t_{\frac{\alpha}{2}} \right|$, reject h_0

Problem Solving Example:

 Given the following pairs of measurements for the two variables:

X	5	8	3	9	10	12
Y	9	12	5	15	18	20

(a) What is the 90% confidence interval for the Y value for $X_0 = 4$?

(b) Find the 95% confidence interval.

(a) The equation is $\hat{y} = a_0 + a_1 x$

$\hat{y} = .007 + 1.681\,x$, so $\hat{y} = .007 + (1.681)(4) = 6.731$ for $x_0 = 4$.

$s_y = .989$, $n = 6$, $\Sigma x = 47$, $\bar{x} = 7.833$, $\Sigma x^2 = 423$, $t_{\alpha/2} = t_{.05}$ with 4 degrees of freedom = 2.132

Then, $E = (2.132)(.989)\sqrt{1 + \dfrac{1}{6} + \dfrac{6(3.833)^2}{6(423) - 2,209}}$

This simplifies to $(2.132)(.989)(1.19775) \approx 2.5255$. Thus, there is a 90% probability that the actual y value is between 4.2055 and 9.2565.

b) Only the $t_{\alpha/2}$ changes its value. $t_{\alpha/2} = t_{.025}$ with 4 degrees of freedom = 2.776. Then, $E = (2.776)(.989)(1.19775) \approx 3.2884$. The required confidence interval becomes $3.4426 < y < 10.0194$.

12.3 Correlation

Regression or estimation enables us to estimate one variable (the dependent variable) from one or more independent variables.

Correlation establishes the degree of relationship between variables. It answers the question: how well does a given equation describe or explain the relationship between independent and dependent variables?

12.3.1 Perfect Correlation

If all values of the variables fit the equation without errors, we say that the variables are perfectly correlated.

The area of a square S is in perfect correlation to its side d that

$$S_9 = d^2.$$

Tossing two coins, we record the result for each coin. Assuming that the coins are fair, there is no relationship between the results for each coin, that is, they are uncorrelated.

Between perfectly correlated and uncorrelated situations, there are situations with some degree of correlation. The heights and weights of people show some correlation. We let x represent one variable (height) and let y represent the other variable (weight). We then try to determine the correlation between x and y.

Simple correlation and simple regression occur when only two variables are involved. When more than two variables are involved, we speak of multiple correlation.

12.3.2 Correlation Coefficient

The degree of relationship between two variables, x and y, is described by the correlation coefficient. If n observations are given

$$(x_i, y_i)\ i = 1, 2, ..., n$$

we can compute the sample correlation coefficient r.

$$r = \frac{\sum (x - \bar{x})(y - \bar{y})}{\sqrt{\sum (x - \bar{x})^2 \sum (y - \bar{y})^2}}$$

We shall list some properties of r:

1. $-1 \leq r \leq 1$

2. $r > 0$ indicates a positive linear relationship, and $r < 0$ indicates a negative linear relationship.

3. $r = 0$ indicates no linear relationship.

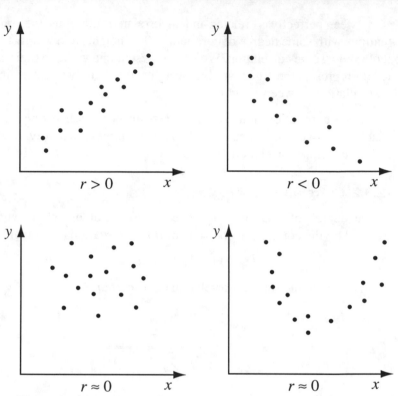

If y tends to increase as x increases, the correlation is called positive.

Consider the linear model

$$y = a_0 + a_1 x + \varepsilon.$$

The total variability of the y's about their mean y can be expressed as

$$\sum (y - \bar{y})^2 = \sum (y - \hat{y})^2 + \sum (\hat{y} - \bar{y})^2$$

$\sum (\hat{y} - \bar{y})^2$ is that portion of the total variability that can be accounted for by the independent variable x. We have

$$\sum (y - \bar{y})^2 = \sum y^2 - \frac{\left(\sum \bar{y} \right)^2}{n}$$

$$\sum (y - \hat{y})^2 = \sum (y - \bar{y})^2 - \frac{\left(\sum (x - \bar{x})(y - \bar{y}) \right)^2}{\sum (x - \bar{x})^2}$$

Hence,

$$\sum (\hat{y} - \bar{y})^2 = \frac{\left(\sum (x - \bar{x})(y - \bar{y}) \right)^2}{\sum (x - \bar{x})^2}$$

and

$$\frac{\sum (y - \hat{y})^2}{\sum (y - \bar{y})^2} = 1 - r^2$$

$$\frac{\sum (\hat{y} - \bar{y})^2}{\sum (y - \bar{y})^2} = r^2$$

The total variation of y defined as

$$\sum (y - \bar{y})^2$$

is the sum of

$$\sum (y - \bar{y})^2 = \sum (y - \hat{y})^2 + \sum (\hat{y} - \bar{y})^2$$

where $\sum (y - \hat{y})^2$ is called the unexplained variation and $\sum (\hat{y} - \bar{y})^2$ is called the explained variation. The deviations $y - \hat{y}$ behave in a random manner, while $\hat{y} - \bar{y}$ have a definite pattern.

We define

$$\text{Coefficient of determination} = \frac{\text{explained variation}}{\text{total variation}}.$$

The value of coefficient of determination lies between zero and one. Note that

and $\qquad r^2 = $ coefficient of determination

$$r = \pm \sqrt{\frac{\text{explained variation}}{\text{total variation}}}.$$

12.3.3 Standard Error of Estimate

A set of measurements is given

$$(x_1, y_1)\ (x_2, y_2),\ \dots\ ,\ (x_n, y_n).$$

Suppose that x is an independent variable. Then, the least-squares regression line of y on x is

$$y = a_0 + a_1 x. \tag{1}$$

Coefficients a_0 and a_1 can be computed from the normal equations

$$\left. \begin{array}{l} \sum y = a_0 n + a_1 \sum x \\[2mm] \sum xy = a_0 \sum x + a_1 \sum x^2 \end{array} \right\} \tag{2}$$

$$a_0 = \frac{(\sum y)(\sum x^2) - (\sum x)(\sum xy)}{n \sum x^2 - (\sum x)^2}$$

$$a_1 = \frac{n \sum xy - (\sum x)(\sum y)}{n \sum x^2 - (\sum x)^2} \tag{3}$$

If y is an independent variable, then

$$x = b_0 + b_1 y \tag{4}$$

and the system of normal equations is

$$\left.\begin{array}{c} \sum x = b_0 n + b_1 \sum y \\ \sum xy = b_0 \sum y + b_1 \sum y^2 \end{array}\right\} \qquad (5)$$

From (5) we find the coefficients

$$b_0 = \frac{(\sum x)(\sum y^2) - (\sum y)(\sum xy)}{n \sum y^2 - (\sum y)^2}$$

$$b_1 = \frac{n \sum xy - (\sum x)(\sum y)}{n \sum y^2 - (\sum y)^2} \qquad (6)$$

By $\hat{y}$, we denote the value of y for a given value of x computed from Equation (1).

We define the standard error of estimate of y on x by

$$s_y = \sqrt{\frac{\sum (y - \hat{y})^2}{n - 2}} \qquad (7)$$

On the other hand, if Equation (4) is used, we have

$$s_x = \sqrt{\frac{\sum (y - \hat{x})^2}{n - 2}} \qquad (8)$$

The values of $\hat{y}$ estimated from the regression line are given by

$$\hat{y} = a_0 + a_1 x. \qquad (9)$$

Hence, (7) can be written as

$$s_y^2 = \frac{\sum (y - \hat{y})^2}{n-2} = \frac{\sum (y - a_0 - a_1 x)^2}{n-2} \tag{10}$$

$$= \frac{\sum y(y - a_0 - a_1 x) - a_0 \sum (y - a_0 - a_1 x) - a_0 \sum x(y - a_0 - a_1 x)}{n-2}$$

We can reduce (10) using

$$\sum (y - a_0 - a_1 x) = \sum y - a_0 n - a_1 \sum x = 0$$

$$\sum x(y - a_0 - a_1 x) = \sum xy - a_0 \sum x - a_1 \sum x^2 = 0$$

since

$$\sum xy = a_0 \sum x = a_1 \sum x^2.$$

Equation (10) becomes

$$s_y^2 = \frac{\sum y(y - a_0 - a_1 x)}{n-2} = \frac{\sum y^2 - a_0 \sum y - a_1 \sum xy}{n-2}. \tag{11}$$

We have defined the total variation as

$$\text{Total variation} = \sum (y - \bar{y})^2$$

and have shown that

Total Variation = Unexplained Variation + Explained Variation

$$\sum (y - \bar{y})^2 = \sum (y - \hat{y})^2 + \sum (\hat{y} - \bar{y})^2$$

The ratio

$$0 \leq \frac{\text{explained variation}}{\text{total variation}} \leq 1$$

is always non-negative. Thus, we have denoted it by r^2.

$$\text{Coefficient of determination} = r^2 = \frac{\text{explained variation}}{\text{total variation}}$$

$$= \frac{\sum(\hat{y} - \bar{y})^2}{\sum(y - \bar{y})^2}$$

Also, we have defined

$$\text{Coefficient of correlation} = r = \pm\sqrt{\frac{\text{explained variation}}{\text{total variation}}}$$

$$= \pm\sqrt{\frac{\sum(\hat{y} - \bar{y})^2}{\sum(y - \bar{y})^2}}$$

Note that r is dimensionless and

$$-1 \le r \le 1$$

The $+$ sign is used for positive linear correlation, and $-$ is used for negative linear correlation.

The standard deviation of y is

$$s = \sqrt{\frac{\sum(y - \bar{y})^2}{n}}. \tag{12}$$

Hence, using (10) we obtain

$$r = \pm\sqrt{\frac{\sum(\hat{y} - \bar{y})^2}{\sum(y - \bar{y})^2}} = \pm\sqrt{\frac{\sum(y - \bar{y})^2 - \sum(y - \hat{y})^2}{\sum(y - \bar{y})^2}} \tag{13}$$

$$= \pm\sqrt{1 - \frac{s_y^2}{s^2}}$$

or

$$s_y = s\sqrt{1 - r^2}. \tag{14}$$

12.3.4 Definition

The equation

$$r = \pm \sqrt{\frac{\text{explained variation}}{\text{total variation}}}$$

is general and applies to linear as well as nonlinear relationships. For the nonlinear case, the values of $\hat{y}$ are computed from an applicable nonlinear equation.

Suppose that the situation is described by

$$y = a_0 + a_1 x + \ldots + a_k x^k \qquad (15)$$

Then, s_y of Equation (11) becomes

$$
s_y = \sqrt{\frac{\sum (y - \hat{y})^2}{n}}
$$
$$
= \sqrt{\frac{\sum y^2 - a_0 \sum y - a_1 \sum xy - \ldots - a_k \sum x^k y}{n}} \qquad (16)
$$

The modified standard error of estimate is given by

$$\hat{s}_y = \sqrt{\frac{n}{n - k - 1}} S_y. \qquad (17)$$

Here, $n - k - 1$ is the number of degrees of freedom.

We shall briefly discuss how the coefficient of correlation works. Suppose that experimental data are gathered, $(x_1, y_1), \ldots, (x_n, y_n)$. Then, we assume some kind of relationship between x and y. First choice is between linear and non-linear.

If the relationship is non-linear, we have to assume some nonlinear equation relating x to y. Then, depending on what type of relationship we assume, the value of r is calculated. One should keep in mind that r depends on what kind of equation is assumed.

Suppose we assume the linear relationship between x and y and the computed value of r is close to 0. All it means is that there is no linear relationship between x and y.

Then, it is possible that:

1. There is no relationship between x and y.

2. There is a non-linear relationship between x and y.

Problem Solving Example:

Q Given the following pairs of measurements for the two variables:

X	5	8	3	9	10	12
Y	9	12	5	15	18	20

(a) Verify the equation: $\sum(Y-\bar{Y})^2 = \sum(Y-\hat{Y})^2 \sum(\hat{Y}-\bar{Y})^2$

(b) What is the value of r^2, the coefficient determination, for part (a)?

 (a) Since we have already found the $\hat{Y}$ values as 8.412, 13.455, 5.05, 15.136, 16.817, and 20.179, we calculate as follows:

$$\sum(Y-\bar{Y})^2 = (9-13.167)^2 + (12-13.167)^2 + (5-13.167)^2 + (15-13.167)^2 + (18-13.167)^2 + (20-13.167)^2 \approx 158.83$$

$$\sum(Y-\hat{Y})^2 = (9-8.412)^2 + (12-13.455)^2 + (5-5.05)^2 + (15-15.136)^2 + (18-16.817)^2 + (20-20.179)^2 \approx 3.915$$

$$\sum(\hat{Y}-\bar{Y})^2 = (8.412-13.167)^2 + (13.455-13.167)^2 + (5.05-13.167)^2 + (15.136-13.167)^2 + (16.817-13.167)^2 + (20.179-13.167)^2 \approx 154.946$$ (Note that : 3.92 + 154.946 is very close to 158.33. The error is due to rounding.)

(b) $r^2 = \dfrac{\sum (\hat{Y} - \overline{Y})^2}{\sum (Y - \overline{Y})^2} = \dfrac{154.95}{158.3} \approx .97$

12.4 Computational Formulas

The least square regression line is

$$y = a_0 + a_1 x$$

which can be written as

$$Y = \left(\frac{\sum XY}{\sum X^2} \right) X \qquad (18)$$

where

$$\begin{aligned} X &= x - \overline{x} \\ Y &= y - \overline{y}. \end{aligned} \qquad (19)$$

Observe that a least-squares line always passes through the point $(\overline{x}, \overline{y})$. If x is an independent variable

$$y = a_0 + a_1 x$$

and

$$\sum y = a_0 + a_1 \sum x,$$

then dividing both sides by n we find

$$\overline{y} = a_0 + a_1 \overline{x}. \qquad (20)$$

Hence,

$$y - \overline{y} = a_1 (x - \overline{x}). \qquad (21)$$

Now we shall prove (18).

For $X = x - \overline{x}$ and $Y = y - \overline{y}$, Equation (21) becomes

$$Y = a_1 X. \qquad (22)$$

The coefficients are

$$a_1 = \frac{n\sum xy - (\sum x)(\sum y)}{n\sum x^2 - (\sum x)^2}$$

$$= \frac{n\sum (X+\bar{x})(Y+\bar{y}) - \sum (X+\bar{x})\sum (Y+\bar{y})}{n\sum (\bar{x}+X)^2 - (\sum (X+\bar{x}))^2}$$

$$= \frac{n\sum XY + n\bar{y}\sum X + n\bar{x}\sum Y + n^2\bar{x}\,\bar{y} - (\sum X + n\bar{x})(\sum Y + n\bar{y})}{n\sum X^2 + 2n\bar{x}\sum X + n^2\bar{x}^2 - (\sum X + n\bar{x})^2}$$

(23)

But

$$\sum X = \sum (x - \bar{x}) = 0$$
$$\sum Y = 0.$$

Hence,

$$a_0 = \frac{n\sum XY + n^2\,\bar{x}\,\bar{y} - n^2\,\bar{x}\,\bar{y}}{n\sum X^2 + n^2\,\bar{x}^2 - n^2\,\bar{x}^2} = \frac{\sum XY}{\sum X^2}$$

(24)

The least square line is

$$Y = \left(\frac{\sum XY}{\sum X^2}\right) X.$$

(25)

For the linear relationship between x and y, the coefficient of correlation is

$$r = \pm\sqrt{\frac{\sum (\hat{y} - \bar{y})^2}{\sum (y - \bar{y})^2}}.$$

(26)

The least-squares regression line of y on x can be written

$$\hat{y} = a_0 + a_1 x$$

or

$$\hat{Y} = a_1 X$$

where

$$a_1 = \frac{\sum XY}{\sum X^2} \quad \text{and} \quad \hat{Y} = \hat{y} - \overline{y}.$$

Thus,

$$r^2 = \frac{\sum (\hat{y} - \overline{y})^2}{\sum (y - \overline{y})^2} = \frac{\sum \hat{Y}^2}{\sum Y^2} = \frac{\sum a_1^2 X^2}{\sum Y^2}$$

$$= \frac{a_1^2 \sum X^2}{\sum Y^2} = \left(\frac{\sum XY}{\sum X^2} \right)^2 \times \frac{\sum X^2}{\sum Y^2} \tag{27}$$

$$= \frac{\left(\sum XY \right)^2}{\left(\sum X^2 \right)\left(\sum Y^2 \right)}$$

and

$$r = \pm \frac{\sum XY}{\sqrt{\left(\sum X^2 \right)\left(\sum Y^2 \right)}}. \tag{28}$$

Observe that for positive linear correlation $\hat{Y}$ increases as X increases and the value of

$$\frac{\sum XY}{\sqrt{\left(\sum X^2 \right)\left(\sum Y^2 \right)}} \tag{29}$$

is positive. Similarly, for negative linear correlation the value of (29) is negative. We can define the coefficient of linear correlation as

$$r = \frac{\sum XY}{\sqrt{(\sum X^2)(\sum Y^2)}} \qquad (30)$$

which automatically assumes the correct sign. Equation (30) is called the product-moment formula for the coefficient of linear correlation.

EXAMPLE:

The data collected are shown in the table below.

x	1	2	4	7	8	10	11
y	1	3	3	5	7	9	9

We want to determine if the relationship between x and y is of linear type. For that purpose, the value of r has to be calculated.

x	y	$X = x - \bar{x}$	$Y = y - \bar{y}$	X^2	XY	Y^2
1	1	−5.143	−4.286	26.450	22.043	18.370
2	3	−4.143	−2.286	17.164	9.471	5.226
4	3	−2.143	−2.286	4.592	4.899	5.226
7	5	0.857	−0.286	0.734	−0.245	0.082
8	7	1.857	1.714	3.448	3.183	2.938
10	9	3.857	3.714	14.876	14.325	13.794
11	9	4.857	3.714	23.59	18.039	13.794

$\sum x = 43$ $\sum y = 37$ $\sum x^2 = 90.86$ $\sum XY = 71.71$ $\sum Y^2 = 59.43$

$$\bar{x} = 6.143$$
$$\bar{y} = 5.286$$

$$r = \frac{\sum XY}{\sqrt{(\sum X^2)(\sum Y^2)}} = \frac{71.71}{\sqrt{90.86 \times 59.43}} = 0.976$$

There is strong positive linear correlation between x and y.

12.4.1 Covariance

The product-moment formula states that

$$r = \frac{\sum XY}{\sqrt{(\sum X^2)(\sum Y^2)}}. \qquad (31)$$

The standard deviations of the variables x and y are

$$s_x = \sqrt{\frac{\sum X^2}{n}} \text{ and} \qquad (32)$$
$$s_y = \sqrt{\frac{\sum Y^2}{n}}.$$

We define the covariance of x and y as

$$s_{xy} = \frac{\sum XY}{n} \qquad (33)$$

The coefficient of correlation can be written as

$$r = \frac{s_{xy}}{s_x s_y} \qquad (34)$$

We shall derive another formula for the linear correlation coefficient.

$$X = x - \bar{x}$$
$$Y = y - \bar{y}$$

$$r = \frac{\sum XY}{\sqrt{(\sum X^2)(\sum Y^2)}} = \frac{\sum (x - \bar{x})(y - \bar{y})}{\sqrt{(\sum (x - \bar{x})^2)(\sum y - \bar{y})^2)}} \qquad (35)$$

But

$$\sum (x - \bar{x})(y - \bar{y}) = \sum xy - \frac{(\sum x)(\sum y)}{n} \qquad (36)$$

and

$$\sum (x - \bar{x})^2 = \sum x^2 - \frac{(\sum x)^2}{n} \qquad (37)$$

$$\sum (y - \bar{y})^2 = \sum y^2 - \frac{(\sum y)^2}{n}. \qquad (38)$$

Substituting (36), (37), and (38) into (35), we obtain

$$r = \frac{n\sum xy - (\sum x)(\sum y)}{\sqrt{(n\sum x^2 - (\sum x)^2)(n\sum y^2 - (\sum y)^2)}}. \qquad (39)$$

12.4.2 Correlation Coefficient for Grouped Data

For grouped data, the values of variables x and y coincide with the corresponding class marks, and the frequencies f_x and f_y are the corresponding class frequencies. By f, we denote the cell frequencies corresponding to the pairs of class marks (x, y).

EXAMPLE:

The weights and ages of a sample of students were recorded. The results are shown in the table.

	Weight		
Age	120-129	130-139	140-149
16-18	2	3	1
19-21	1	5	0
22-24	3	6	2
25-28	2	5	4

cell frequency

Consider the shaded cell. The class marks of this cell are

$$\text{weight} = 134.5$$

$$\text{age} = 20$$

The cell frequency of the cell (134.5, 20) is 5.

Equation (39) becomes

$$r = \frac{n\sum fxy - (\sum f_x x)(\sum fyy)}{\sqrt{[n\sum f_x x^2 - (\sum f_x x)^2][n\sum f_y y^2 - (\sum f_y y)^2]}}. \tag{40}$$

This method is called the cooling method.

12.4.3 Linear Correlation Coefficient

Regression line of y on x is

$$y = a_0 + a_1 x$$

or

$$y - \overline{y} = \frac{rs_y}{s_x}(x - \overline{x}) \tag{41}$$

or

$$Y = \frac{rs_y}{s_x} X.$$

If we consider y to be an independent variable, then

$$x = b_0 + b_1 y$$

or

$$x - \bar{x} = \frac{rs_x}{s_y}(y - \bar{y}) \qquad (42)$$

or

$$X = \frac{rs_x}{s_y} Y.$$

If $r = \pm 1$, then lines (41) and (42) are identical and there is perfect linear correlation between x and y. The slopes of lines (41) and (42) are equal if and only if $r = \pm 1$.

If $r = 0$, then there is no linear correlation and the lines are at right angles.

12.4.4 Spearman's Formula for Rank Correlation

Sometimes, the exact values of the variables are impossible or impractical to calculate. In such cases, we use the method of ranking the data according to their importance, size, quality, etc. using the numbers $1, 2, \ldots, n$. For two ranked variables x and y we define the coefficient of rank correlation

$$r = 1 - \frac{6\sum d^2}{n(n^2 - 1)}$$

where n is the number of all pairs (x, y) of ranked data and d is the difference between ranks of corresponding values of x and y.

12.4.5 Sampling

Suppose the experimental situation requires the measurement of two variables x and y. The results are assembled in pairs

$$(x_1, y_1), \ldots, (x_n, y_n).$$

These pairs constitute a sample from a population of all possible pairs. Populations that involve two variables are called the bivariate populations. We can compute the coefficient of correlation r for the sample (x_1, y_1), ..., (x_n, y_n). Based on it, we can make estimates concerning the coefficient of correlation R for the whole population.

To test the significance or hypotheses of various values of R, we have to know the sampling distribution of r.

12.4.6 Test of Hypothesis R = 0

For $R = 0$, this distribution is symmetrical and we can use a statistic involving Student's "t" distribution.

The statistic

$$t = \frac{r\sqrt{n-2}}{\sqrt{1-r^2}}$$

has Student's "t" distribution for $n - 2$ degrees of freedom.

12.4.7 Test of Hypothesis R = R₀ ≠ 0

For $R \neq 0$, the sampling distribution is skewed. The Fisher's Z transformation yields a statistic that is very close to normal distribution:

$$Z = \frac{1}{2} ln\left(\frac{1+r}{1-r}\right) = 1.1513 \log_{10}\left(\frac{1+r}{1-r}\right).$$

The statistic Z has mean

$$u_Z = \frac{1}{2} ln\left(\frac{1+R_0}{1-R_0}\right) = 1.1513 \log_{10}\left(\frac{1+R_0}{1-R_0}\right)$$

and standard deviation given by

$$\sigma_Z = \frac{1}{\sqrt{n-3}}.$$

Suppose that two samples of sizes n_1 and n_2 are drawn from the population. The correlation coefficient for samples n_1 and n_2 are r_1 and r_2, respectively.

Using r_1 and r_2, we compute Z_1 and Z_2

$$Z_1 = \frac{1}{2} \ln \left(\frac{1 + r_1}{1 - r_1} \right)$$

$$Z_2 = \frac{1}{2} \ln \left(\frac{1 + r_2}{1 - r_2} \right).$$

The test statistic

$$Z = \frac{Z_1 - Z_2 - \mu_{Z_1 - Z_2}}{\sigma_{Z_1 - Z_2}}$$

where

$$\mu_{Z_1 - Z_2} = \mu_{Z_1} - \mu_{Z_2}$$

and

$$\sigma_{Z_1 - Z_2} = \sqrt{\sigma_{Z_1}^2 + \sigma_{Z_2}^2} = \sqrt{\frac{1}{n_1 - 3} + \frac{1}{n_2 - 3}}$$

has normal distribution. Thus, we can decide if the two correlation coefficients r_1 and r_2 differ significantly from each other.

Problem Solving Examples:

Q "There is a complex system of relationships in the business world. As an example, the number of new movies that appears in the course of a week has an appreciable effect on the weekly change in the Dow-Jones Industrial Average."

This is the opinion of a certain armchair economist. This fellow hires you as a consultant and expects you to test his theory. In the first five weeks you observe the following:

X, number of new movies	1	2	4	5	5
Y, change in Dow-Jones Industrial Average	–2	4	–5	7	–8

What is the correlation between X and Y? What implications does this have for the theory?

 To compute the correlation coefficient r, we construct the following table:

X	Y	X^2	Y^2	XY
1	–2	1	4	–2
2	4	4	16	8
4	–5	16	25	–20
5	7	25	49	35
5	–8	25	64	–40

$n = 5$

$$\sum X = 17 \qquad \sum Y = -4 \qquad \sum X^2 = 71$$

$$\sum Y^2 = 158 \qquad \sum XY = -19$$

$$
\begin{aligned}
r &= \frac{n\sum XY - \sum X \sum Y}{\sqrt{n\sum X^2 - \left(\sum X\right)^2}\sqrt{n\sum Y^2 - \left(\sum Y\right)^2}} \\[2mm]
&= \frac{5(-19) - (17)(-4)}{\sqrt{5(71) - (17)^2}\sqrt{5(158) - (-4)^2}} \\[2mm]
&= \frac{-95 + 68}{\sqrt{66}\sqrt{774}} \\[2mm]
&= \frac{-27}{(8.12)(27.82)} = -.119
\end{aligned}
$$

The coefficient of correlation indicates a weak negative relation between the number of new movies and changes in the Dow-Jones Industrial Average. This seems to cast doubt on the theory.

The table below lists the ranks assigned by two securities analysts to 12 investment opportunities in terms of the degree of investor risk involved.

Investment	Rank by Analyst 1	Rank by Analyst 2
A	7	6
B	8	4
C	2	1
D	1	3
E	9	11
F	3	2
G	12	12
H	11	10
I	4	5
J	10	9
K	6	7
L	5	8

Find the correlation between the two rankings. What is the relationship between the two rankings?

We compute r_s, the coefficient of rank correlation between the rankings given by Analyst 1 and Analyst 2.

Investment	Analyst 1	Analyst 2	$X_i - Y_i$	$(X_i - Y_i)^2$
A	7	6	1	1
B	8	4	4	16
C	2	1	1	1
D	1	3	-2	4
E	9	11	-2	4
F	3	2	1	1
G	12	12	0	0
H	11	10	1	1
I	4	5	-1	1
J	10	9	1	1
K	6	7	-1	1
L	5	8	-3	9

$$r_s = 1 - \frac{6 \sum_{i=1}^{12} (X_i - Y_i)^2}{n(n^2 - 1)}$$

$$\sum_{i=1}^{12} (X_i - Y_i)^2 = 40$$

$$r_s = 1 - \frac{6(40)}{12(144 - 1)} = 1 - \frac{240}{1,716}$$

$$= 1 - .14$$

$$r_s = .86$$

The correlation between the rankings given by Analyst 1 and Analyst 2 is .86. This is a very strong, positive correlation between the two rankings. It seems to imply that these two analysts have very similar ideas about the degree of investor risk involved with the 12 securities.

Q The data in the table below represent the scores on a mathematics placement test and the final averages in a freshman mathematics class. These 11 scores constitute a random sample. Test the usefulness of the placement test in predicting the performance of students in the course.

Test Score X	Final Average in Math Y
51	75
52	72
59	82
45	67
61	75
54	79
56	78
67	82
63	87
53	72
60	96

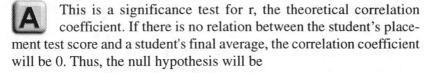 This is a significance test for r, the theoretical correlation coefficient. If there is no relation between the student's placement test score and a student's final average, the correlation coefficient will be 0. Thus, the null hypothesis will be

$$H_0: \rho \le 0.$$

If there is a relationship between the placement test score and the student's final average, we would expect the relationship to be as follows. If a student does well on the placement test score, then we would expect him/her to do well in the course. If a student does poorly on the test, we would expect this same student to do poorly in the course. A relationship of this type between two variables, X and Y, if such a relationship exists, is called positive. The correlation coefficient of positively correlated variables is positive. Thus, the alternative hypothesis is

$$H_A: \rho > 0.$$

It is natural for the test statistic to be r, the sample correlation coefficient, based on the data.

We have already found a way to derive a normally distributed random variable from r by using Fisher's z-transform. Using r, we let

$$z_f = 1.1513 \log_{10} \frac{1+r}{1-r}$$

or

$$z_f = \frac{1}{2} \log_e \frac{1+r}{1-r}.$$

z_f is a normally distributed random variable with mean

$$\mu_z = \frac{1}{2} \log_e \frac{1+\rho}{1-\rho} \quad \text{and variance} \quad \sigma_z^2 = \frac{1}{n-3}.$$

We know that

$$z = \frac{z_f - \mu_z}{\sigma_z}$$

is a standard normal random variable.

We next set a level of significance for our test. We choose a common level of significance $\alpha = .01$.

With the level of significance, we determine the critical value of the test. From the table of the standard normal distribution, we see that a Z-score of 2.33 or more will appear with probability of .01.

$$Pr(Z \geq 2.33) = .01$$

Thus, 2.33 will be the critical value for our test.

To carry out the test, we follow this procedure:

(1) Compute r, the sample correlation coefficient.

(2) Use the table of the Fisher z-transform to compute z_f.

(3) Let $Z = \dfrac{z_f - \mu_z}{\sigma_z}$, this is the calculated z-statistic.

(4) Compare Z with 2.33. If $Z > 2.33$, reject H_0. If $Z < 2.33$, accept H_0.

These steps are followed below:

$$(1) \quad r = \frac{n \sum X_i Y_i - \left(\sum X_i\right)\left(\sum Y_i\right)}{\sqrt{n\left(\sum X_i^2\right) - \left(\sum X_i\right)^2} \times \sqrt{n\left(\sum Y_i^2\right) - \left(\sum Y_i\right)^2}}$$

Test score X	Final average Y	X^2	Y^2	XY
51	75	2,601	5,625	3,825
52	72	2,704	5,184	3,744
59	82	3,481	6,724	4,838
45	67	2,025	4,489	3,015
61	75	3,721	5,625	4,575
54	79	2,916	6,241	4,266
56	78	3,136	6,084	4,368
67	82	4,489	6,724	5,494
63	87	3,969	7,569	5,481
53	72	2,809	5,184	3,816
60	96	3,600	9,216	5,760

$$n = 11$$

$$\sum X_i = 621 \qquad \sum X_i^2 = 35,451$$

$$\sum Y_i = 865 \qquad \sum Y_i^2 = 68,665 \qquad \sum XY = 49,182$$

$$\overline{X} = \frac{621}{11} = 56.45 \qquad \overline{Y} = \frac{865}{11} = 78.64$$

Thus,

$$r = \frac{(49,182) - (11)(56.45)(78.64)}{\sqrt{[35,451 - 11(56.45)^2][68,665 - 11(78.64)^2]}}$$

$$= 49,182 - 48,831.5 = \frac{350.5}{\sqrt{(398.4)(638.3)}} = \frac{350.5}{504.3} = .69$$

(2) We now compute z_f from r. Using the table of Fisher z-values, we see that the z_f that corresponds to $r = .69$ is .848 (correct to three decimal places).

(3) Under the null hypothesis,

$$\mu_z = \frac{1}{2} \log_e \frac{1 + \rho}{1 - \rho} = \frac{1}{2} \log_e \frac{1 + 0}{1 - 0}$$

$$= \frac{1}{2} \log_e 1 = 0$$

There are 11 observations, thus

$$\sigma_z = \sqrt{\sigma_z^2} = \sqrt{\frac{1}{n - 3}}$$

$$= \sqrt{\frac{1}{11 - 3}} = \frac{1}{\sqrt{8}}.$$

The calculated Z-statistic is

$$Z = \frac{z_f - \mu_z}{\sigma_z} = \frac{.848 - 0}{\frac{1}{\sqrt{8}}}$$

$$= \left(\sqrt{8}\right)(.848)$$

$$= 2.398$$

$$= 2.4.$$

(4) We compare the Z-statistic with the critical value and see that $2.4 > 2.33$. Therefore, we reject the null hypothesis $\rho \leq 0$ in favor of the alternative hypothesis $\rho > 0$.

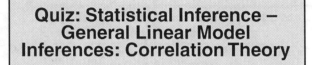

Quiz: Statistical Inference – General Linear Model Inferences: Correlation Theory

1. A drug company has published results of medical tests that claim that a certain medication is effective in 60% of the cases of a rare disease. A medical researcher questions this result, although she does not know whether she thinks it is high or low. She only has 12 people to test with the medication. What test statistic, or key value of t, should she check her results against if she wants to limit the possibility of a Type I error to 1.0% at the most?

    (A) 2.681   (D) 3.055

    (B) 2.764   (E) 4.025

    (C) 3.106

2. The test scores of eight students are compared before and after a new computerized teaching method is put into effect. What is the critical value of t that is used to determine if the new method had the effect of increasing scores at the 0.05 level of significance when you subtract the before scores from the after scores for each student?

    (A) −1.860   (D) +2.365

    (B) +1.895   (E) −2.365

    (C) −1.895

3. Five hundred applicants took a college entrance examination. You have 10 minutes to present a summary of the applicants' test performance. The computer has suddenly broken down. You must estimate the mean score of the candidates from a random sample of 10, equipped with only a pocket calculator. Here are the test scores of the ten candidates you selected:

    20, 14, 29, 26, 10, 34, 20, 30, 17, and 20.

    Compute the 90% confidence interval for the mean score of the 500 applicants.

    (A) 17.6 to 26.4   (D) 7.587 to 22

    (B) 17.6 to 22   (E) None of these

    (C) 22 to 23.4

4. In finding the relationship between two variables, the equation of the regression line is determined to be:

    $$\hat{y} = 250 - 3x.$$

    The five values of the independent variable used in the experiment were: 2, 5, 8, 9, and 11. What is $\bar{y}$?

    (A) 250   (D) 206

    (B) 247   (E) Cannot be determined

    (C) 229

5. When two variables are investigated for a relationship, it is discovered that the greatest absolute value of r is attained when the data is fit to a regression curve following the equation

    $$y = ax^b.$$

    What is this kind of regression called?

    (A) Exponential regression   (D) Quartic regression

    (B) Power regression   (E) Logarithmic regression

    (C) Linear regression

(1)   (2)

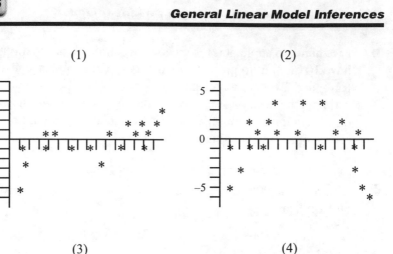

(3)   (4)

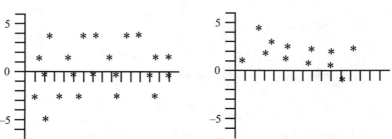

6.   Which of the graphs above suggest approximately linear relation-
     ships between the dependent and the independent variables?

     (A)   1 and 2   (D)   1 and 4

     (B)   2 and 3   (E)   2 and 4

     (C)   3 and 4

7.   The value of t in the t-test for the correlation coefficient is given
     by the formula

$$t = r\sqrt{\frac{n-2}{1-r^2}}$$

     with the number of degrees of freedom given by $n - 2$. Six pairs
     of independent and dependent variables are measured in a two-

tailed situation, and a value for the correlation coefficient of 0.8 is determined. For this data, what is the minimum value μ, of the level of significance, at which you could reject the null hypothesis? (The null hypothesis states that there is no significant relationship between the independent and dependent variables.)

(A) 0.20

(D) 0.02

(B) 0.10

(E) 0.01

(C) 0.05

8. When attempting to show the relationship between two variables, a regression line can be drawn and the correlation coefficient, r, can be calculated. The y-intercept and slope of the regression are denoted b_0 and b_1, respectively. Consider the following statements:

I. $r = b_0$
II. $r = b_1$
III. r and b_1 have the same sign.

Which of the statements above is always true?

(A) I only

(B) II only

(C) III only

(D) I and III only

(E) None of the statements is always true.

9. Two variables are studied and the correlation coefficient, r, is determined to have the value −0.05. Consider the following statements about the relationship between the variables. Which statements are correct?

I. There is a strong relationship.
II. There is a weak relationship.
III. As one increases, the other increases.
IV. As one increases, the other decreases.
V. A casual connection between variables has been proven.

(A) I and II only   (D) II and IV only

(B) II and III only   (E) I, II, III, and V only

(C) I and IV only

10. We compare the two regression lines representing the relationship between two variables, Graph M and Graph N. Consider the following statements relating the correlation coefficient for the two representations:

I. r_M and r_N are both negative.
II. $|r_M| < |r_N|$.
III. $r_M < r_N$.

   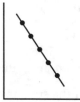

Graph M   Graph N

Which of the statements is correct?

(A) I only   (D) I and II only

(B) II only   (E) All of them

(C) III only

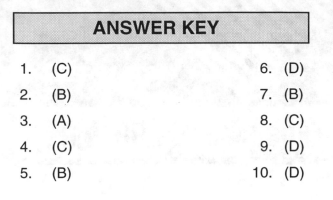

ANSWER KEY

1.	(C)	6.	(D)
2.	(B)	7.	(B)
3.	(A)	8.	(C)
4.	(C)	9.	(D)
5.	(B)	10.	(D)

CHAPTER 13

Experimental Design

13.1   Design of an Experiment

A statistician starts his/her job with the design of an experiment. Depending on the experiment, he/she needs a certain amount of information. The cost of gathering information should also be considered.

The design of an experiment should be cost effective, and it should guarantee that the necessary information will be supplied.

Each design of an experiment consists of the following steps:

1.   Statement of an objective. It usually consists of a description of the population and a description of the parameters of the population to be estimated.

2.   Statement of the amount of information required about the parameters.

3.   Experimental design which consists of selection of the appropriate experimental plan.

4.   Estimation of test procedure.

Suppose that the objective is to find the parameter μ of the population. We have to decide how accurate μ should be. For that purpose, we can apply a bound on the error of the estimate of μ. That bound can be $\pm A$ units from μ.

Note that A can assume any value we choose.

Suppose that μ is a mean yearly income of a construction worker. Then, we can choose $A = \$1.00$ and estimate μ to be within $\$1.00$.

Often, the sample mean is used as a point estimate of μ; then, the bound on the error of estimate is

$$\frac{2\sigma}{\sqrt{n}}.$$

Solving the equation

$$A = \frac{2\sigma}{\sqrt{n}}$$

for n

$$n = \frac{u\sigma^2}{A^2}$$

we find the sample size required to estimate μ to within A units.

Two terms are used frequently in descriptions of experimental design.

1. **Experimental Unit** – Any object or person on which a measurement is made.

2. **Treatment** – A factor level, or combination of factor levels, applied to an experimental unit.

Suppose that our experimental objective is to estimate the parameters μ and $\mu_1 - \mu_2$, or to test a hypothesis about them.

It is important to determine the quantity of information in an experiment relative to the parameter.

Often, we estimate a parameter, λ, using a point estimate $\hat{\lambda}$. For the sampling distribution of the point estimate, which is approximately normal with $\hat{\lambda}$ and standard deviation $\sigma_{\hat{\lambda}}$, the bound on the error of estimate is $2\sigma_{\hat{\lambda}}$.

If we are looking for a point estimate of λ, then we can set the bound on the error of the estimate, say A.

In general,

$$\sigma_{\hat{\lambda}} = \sigma_{\hat{\lambda}}(n)$$

that is, $\sigma_{\hat{\lambda}}$ depends on n. Hence, solving the equation

$$A = 2\sigma_{\hat{\lambda}}$$

for n we find the sample size n necessary to achieve the bound on the error equal to A.

We summarize the results concerning sample sizes in the table.

λ	$\hat{\lambda}$	$\sigma_{\hat{\lambda}}$	Sample Size
μ	$\bar{y}_1$	$\dfrac{\sigma}{\sqrt{n}}$	$n = \dfrac{\mu\sigma^2}{A^2}$
$\mu_1 - \mu_2$	$\bar{y}_1 - \bar{y}_2$	$\sqrt{\dfrac{\sigma_1^2}{n_1} + \dfrac{\sigma_2^2}{n_2}}$	$n = \dfrac{\mu(\sigma_1^2 + \sigma_2^2)}{A^2}$

Note that in estimating

$$\mu_1 - \mu_2$$

we use

$$\bar{y}_1 - \bar{y}_2$$

and determine the sample size by solving the equation

$$A = 2\sqrt{\dfrac{\sigma_1^2}{n_1} + \dfrac{\sigma_2^2}{n_2}}.$$

We want to estimate $\mu_1 - \mu_2$ to within A units.

EXAMPLE:

A chemist wants to estimate the difference in mean melting temperatures for two different alloys. From previous experiments, he knows that the range in melting temperatures for each alloy is approximately 400°. How many independent random samples of each kind of alloy must be checked to estimate

$$T_1 - T_2$$

within 50°?

The range of melting temperatures for both alloys is the same; thus, we can assume that the population variances σ_1^2 and σ_2^2 are approximately the same.

$$\sigma_1^2 = \sigma_2^2 = \sigma^2$$

The range estimate of σ is

$$\hat{\sigma} = \frac{\text{range}}{4} = \frac{400}{4} = 100.$$

The sample size formula is

$$n = \frac{4(\sigma_1^2 + \sigma_2^2)}{A^2}$$

$$= \frac{4(100^2 + 100^2)}{50^2} = 32.$$

Thus, we should examine 32 samples of each type of alloy to estimate $T_1 - T_2$ with a bound on the error of estimate $A = 50°$.

Observe that by increasing the desired bound on the error of estimate, we decrease the number of required samples, i.e., decrease the cost of conducting the experiment.

For example, for

$$A = 75°$$

$$n = \frac{4(100^2 + 100^2)}{75^2} = 14.$$

Now, suppose the objective is to find an interval estimate of the parameter λ. Let $\hat{\lambda}$ be a point estimate of the parameter λ. We assume that the sampling distribution of the point estimates is approximately normal, with mean λ and standard deviation $\sigma_{\hat{\lambda}}$.

For confidence coefficient

$$1 - \alpha$$

the confidence interval for λ is

$$\hat{\lambda} \pm Z_{\frac{\alpha}{2}} \sigma_{\hat{\lambda}}.$$

One of the accepted measures of the amount of information important to the parameter λ is the half width $Z_{\frac{\alpha}{2}} \sigma_{\hat{\lambda}}$ of the confidence interval.

$$A = Z_{\frac{\alpha}{2}} \sigma_{\hat{\lambda}}$$

Solving this equation for n, we find the sample size required to estimate a parameter λ by using the confidence coefficient

$$1 - \alpha$$

and a confidence interval $\pm Z_{\frac{\alpha}{2}} \sigma_{\hat{\lambda}}$.

The table that follows shows the results of interval estimates of the parameters μ and $\mu_1 - \mu_2$.

λ	$\hat{\lambda}$	Confidence Interval	Sample Size
μ	$\bar{y}$	$\bar{y} \pm Z_{\frac{\alpha}{2}} \dfrac{\sigma}{\sqrt{n}}$	$n = \dfrac{Z_{\frac{\alpha}{2}}^2 \sigma^2}{A^2}$
$\mu_1 - \mu_2$	$\bar{y}_1 - \bar{y}_2$	$(\bar{y}_1 - \bar{y}_2) \pm Z_{\frac{\alpha}{2}} \sqrt{\dfrac{\sigma_1^2}{n} + \dfrac{\sigma_2^2}{n}}$	$n = \dfrac{Z_{\frac{\alpha}{2}}^2 \left(\sigma_1^2 + \sigma_2^2 \right)}{A^2}$

Suppose that we want to test the research hypothesis

$$h_a : \lambda > \lambda_0.$$

Then, the null hypothesis is

$$h_0 : \lambda = \lambda_0$$

Assume that the distribution of $\hat{\lambda}$ is approximately normal with mean λ_0 and standard deviation $\sigma_{\hat{\lambda}}$ under the null hypothesis. We want the probability of a Type 1 error to be α and the probability of a Type 2 error to be β or less for the actual value of λ such that

$$\lambda - \lambda_0 \geq \Delta$$

i.e., λ lies a distance of Δ or more above λ_0.

λ	h_0	Δ	Test Statistic	Sample Size
μ	$\mu = \mu_0$	$\mid \mu - \mu_0 \mid$	$Z = \dfrac{\bar{y} - \mu_0}{\sigma} \sqrt{n}$	$n = \dfrac{\sigma^2 (Z_\alpha + Z_\beta)^2}{\Delta^2}$
$\mu_1 - \mu_2$	$\mu_1 - \mu_2 = \delta$	$\mid \mu_1 - \mu_2 - \delta \mid$	$Z = \dfrac{(\bar{y}_1 - \bar{y}_2) - \delta}{\sqrt{\dfrac{\sigma_1^2}{n} + \dfrac{\sigma_2^2}{n}}}$	$n = \dfrac{(\sigma_1^2 + \sigma_2^2)(Z_\alpha + Z_\beta)^2}{\Delta^2}$

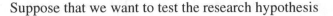

Problem Solving Examples:

Q Suppose that light bulbs made by a standard process have an average life of 2,000 hours, with a standard deviation of 250 hours, and suppose that it is worthwhile to change the process if the mean life can be increased by at least 10 percent. An engineer wishes to test a proposed new process, and he is willing to assume that the standard deviation of the distribution of lives is about the same as for the standard process. How large a sample should he examine if he wishes the probability to be about .01 that he will fail to adopt the new process if in fact it produces bulbs with a mean life of 2,250 hours?

A For the engineer to adopt the process, the new mean must be a 10% increase over the old. Since 10% of 2,000 is 200, the new sample mean would have to be at least 2,200. Since we know that the true mean is 2,250, then $\overline{X} - \mu$ would have to be less than −50 in order to reject it. We want $Pr(\overline{X} - m < -50) = .01$. Divide by

$\sigma / \sqrt{n}$ to obtain $\Pr\left[(\overline{x} - \mu)/(\sigma / \sqrt{n}) < -50/(\sigma / \sqrt{n})\right] = .01$.

By the Central Limit Theorem, $(\overline{x} - \mu)/(\sigma / \sqrt{n})$ is a standard normal quantity, and thus can be replaced by z.

Hence, we have $\Pr\left[z < -50//(\sigma / \sqrt{n})\right] = .01$

By the standard normal table, $Pr(Z < -2.33) = .01$

Then $-50/(\sigma / \sqrt{n}) = -2.33$. Using $s = 250$ as an approximation for σ, we have $-50/(250 / \sqrt{n}) = -2.33$. But since $s = 250$

This implies $\sqrt{n} \approx 11.65$, so $n \approx 135.7225$. Thus, the sample should include at least 136 observations.

Q A research worker wishes to estimate the mean of a population using a sample large enough that the probability will be .95 that the sample mean will not differ from the population mean by more than 25% of the standard deviation. How large a sample should he take?

A Twenty five percent of the standard deviation is $\dfrac{1}{4}$ of it, $\dfrac{\sigma}{4}$. We want

$$\Pr\left[\,|\overline{X} - \mu| < \frac{\sigma}{4}\right] = .95.$$

Equivalently, $\Pr\left[-\dfrac{\sigma}{4} < X - \mu < \dfrac{\sigma}{4}\right] = .95.$

Divide through by $\dfrac{\sigma}{\sqrt{n}}$:

$$\Pr\left[\left|(\bar{x} - \mu)\right|/\left(\sigma/\sqrt{n}\right) < (\sigma/4)\left(\sigma/\sqrt{n}\right)\right] = .95$$

This translates to

$$\Pr\left[-(\sigma/4)/\left(\sigma/\sqrt{n}\right) < z < (\sigma/4)/\left(\sigma/\sqrt{n}\right)\right] \text{ or}$$

$$\Pr\left[-\sqrt{n}/4 < z < \sqrt{n}/4\right]$$

From the standard normal table, $\Pr[-1.96 < z < 1.96] = .95$

Then $\sqrt{n}/4 = 1.96$

Solving, $\sqrt{n} = 7.84$ and $n = 61.4656$

Thus, the sample should include at least 62 observations. (For this type of problem, estimates should be rounded UP to the nearest whole number.)

 Find a 90% confidence interval for $\mu_1 - \mu_2$ when $n = 10$, $m = 7$, $\bar{X} = 4.2$, $\bar{Y} = 3.4$, $S_1^2 = 49$, and $S_2^2 = 32$.

 First we look at the t table and find
$$Pr(-1.753 < t_{(m+n-2 = 17 - 2 = 15)} < 1.753) = .90$$
We have derived a confidence interval for $\mu_1 - \mu_2$ and found it to be

$$(\bar{X} - \bar{Y}) - t_{\frac{\alpha}{2}}\sqrt{\dfrac{(n-1)S_1^2 + (m-1)S_2^2}{n+m-2}\left(\dfrac{1}{n} + \dfrac{1}{m}\right)},$$

$$(\bar{X} - \bar{Y}) + t_{\frac{\alpha}{2}}\sqrt{\dfrac{(n-1)S_1^2 + (m-1)S_2^2}{n+m-2}\left(\dfrac{1}{n} + \dfrac{1}{m}\right)}.$$

Now $t_{\frac{\alpha}{2}}$ = 1.753. Let us insert our values

$$(4.2-3.4)-1.753\sqrt{\frac{(10-1)49+(7-1)32}{10+7-2}\left(\frac{1}{10}+\frac{1}{10}\right)},$$

$$(4.2-3.4)+1.753\sqrt{\frac{(10-1)49+(7-1)32}{10+7-2}\left(\frac{1}{10}+\frac{1}{10}\right)},$$

Combining, we obtain $0.8-1.753\sqrt{\frac{9(49)+6(32)}{15}\left(\frac{17}{70}\right)},$

$$0.8+1.753\sqrt{\frac{9(49)+6(32)}{15}\left(\frac{17}{70}\right)},$$

$$\left(0.8-1.753\sqrt{\frac{633}{15}\left(\frac{17}{70}\right)}, 0.8+1.753\sqrt{\frac{633}{15}\left(\frac{17}{70}\right)}\right).$$

Now we convert to decimals.

$$\left[0.8-1.753\sqrt{10.25}, 0.8+1.753\sqrt{10.25}\right]$$

or $\qquad [0.8-1.753(3.202), 0.8+1.753(3.202)].$

Our final answer is $(-4.813, 6.413)$.

Q The weights of 15 New York models had a sample mean of 107 lbs. and a sample standard deviation of 10 lbs. Twelve Philadelphia models had a mean weight of 112 lbs. and a standard deviation of 8 lbs. Make a .90 confidence interval estimate of the difference of the mean weights between the two model populations.

With small samples and an unknown variance, we can immediately quote our previously derived result. The interval comes from a t statistic and is

$$\left((\overline{X}-\overline{Y})-t_{\frac{\alpha}{2}}\sqrt{\frac{(n-1)S_1^2+(m-1)S_2^2}{n+m-2}\left(\frac{1}{n}+\frac{1}{m}\right)},\right.$$

$$\left.(\overline{X}-\overline{Y})+t_{\frac{\alpha}{2}}\sqrt{\frac{(n-1)S_1^2+(m-1)S_2^2}{n+m-2}\left(\frac{1}{n}+\frac{1}{m}\right)}.\right.$$

The number of degrees of freedom is: $n+m-2=15+12-2=25$. Therefore, b is the value such that

$$Pr(-t_{\frac{\alpha}{2}} < t(25) < t_{\frac{\alpha}{2}}) = .90.$$

From the tables, we see that $t_{\frac{\alpha}{2}} = 1.708$.

The problem tells us $\overline{X} = 107$, $\overline{Y} = 112$, $S_1^2 = 10^2 = 100$, $S_2^2 = 8^2 = 64$, $n = 15$, and $m = 12$. Inserting these values we obtain:

$$\left(-5-1.708\sqrt{\frac{(14)100+11(64)}{25}\left(\frac{27}{180}\right)},\right.$$

$$\left.-5+1.708\sqrt{\frac{(14)100+11(64)}{25}\left(\frac{27}{180}\right)}.\right.$$

Further simplification results in:

$$[-5 - 1.708\,(3.553), -5 + 1.708\,(3.553)].$$

Equivalently, $(-5 - 6.069, -5 + 6.069)$.

Our final confidence interval for the difference in models' weights is $(-11.069, 1.069)$.

13.2   Completely Randomized Design

Sometimes, we deal with more than one population. The methods of estimating a population mean, μ, or the difference between two population means, $\mu_1 - \mu_2$, described in 13.1 have to be extended. The completely randomized design enables us to compare u ($u \geq 2$) population means,

$$\mu_1 , \mu_2 , ..., \mu_u .$$

We assume here that there are u different populations. From these populations, we draw independent random samples of size

$$n_1 , n_2 , ... , n_u$$

respectively.

Thus, we assume that there are

$$n_1 + n_2 + ... + n_u$$

experimental units. Each unit receives a treatment and treatments are randomly assigned to the experimental units in such a way that

n_1 units receive treatment 1

n_2 units receive treatment 2

.
.
.

n_u units receive treatment u

After the corresponding treatment means are calculated, we will make inferences concerning them.

EXAMPLE:

A factory has four different methods at its disposal to test the quality and durability of the bearings it makes. It is important to determine if there is a difference in mean test readings for bearings using four different methods.

Here, the experimental units are bearings and the treatments are four methods of testing $A, B, C,$ and D. A random sample of 16 bearings is chosen, and 4 bearings are randomly assigned to each testing method. We have a completely randomized design with four observations for each treatment.

	Bearing			
Testing Method	A	B	C	D
	A	B	C	D
	A	B	C	D
	A	B	C	D

Suppose that four technicians are assigned to perform tests. We number them 1, 2, 3, and 4. The bearings are randomly assigned, four to each technician.

One of the possible random assignments is shown in the table.

Technician			
1	2	3	4
A	B	C	D
A	B	C	D
A	B	C	D
A	B	C	D

For each treatment, we have four observations. Now, any detected difference may be due to differences among methods or differences among technicians.

Suppose that the hypothesis

$$h_0 : \mu_A - \mu_B = 0$$

was tested against

$$h_a : \mu_A - \mu_B \neq 0$$

The rejection of h_0 could be due to the differences in the methods of testing, or due to the fact that technicians represent various skill levels.

The completely randomized design presented can be modified to gain more reliable information about the means μ_A, μ_B, μ_C, and μ_D. We can impose one restriction upon the random choice of methods of testing.

Each technician can be asked to use each of the methods exactly once. The order of tests for each technician is randomized.

One of the possible designs is shown in the table.

Technician

		1	2	3	4
Day	1	B	C	A	A
	2	D	A	D	D
	3	A	D	B	C
	4	C	B	C	B

This design, which is an improved completely randomized design, is called a randomized block design. Here, the blocks are technicians. The influence of technicians upon the results μ_A, μ_B, μ_C, and μ_D in this design has been eliminated.

Now, the hypothesis

$$h_0 : \mu_A - \mu_B = 0$$

is rejected, and we know that the difference between μ_A and μ_B is due to the differences detected by the methods A and B.

EXAMPLE:

The situation is as described in the previous example. The instruments used to test the bearings require careful calibration and adjustments, and each of the technicians is able to perform only one test a day.

We use the randomized block design table and denote the first row as a workload of day 1, the second row as a workload of day 2, etc.

Technician

		1	2	3	4
Day	1	B	C	A	A
	2	D	A	D	D
	3	A	D	B	C
	4	C	B	C	B

Now, the possible differences in μ_A, μ_B, μ_C, and μ_D can be due to the fact that no D method was used on day 1, while three D methods were used on day 2. Hence, if we reject

$$h_0 : \mu_B - \mu_D = 0$$

we would not be certain if the difference between μ_B and μ_D is due to the difference in testing methods B and D, or if it is due to the difference among days.

In this situation, the experimental units are affected by two foreign sources of variability: technicians and days.

We can take care of this by using the Latin square design. The completely randomized design was modified to eliminate the technician as an extraneous source of variability. The obtained design was called the randomized block design.

We can further modify this design to eliminate the influence of days, that is, the variability among days. The randomization is restricted to guarantee that each method appears in each row and in each column. Such an experimental design is called a Latin square design.

One of possible Latin square designs is shown in the following table.

Technician

		1	2	3	4
Day	1	*B*	*C*	*A*	*D*
	2	*D*	*A*	*B*	*C*
	3	*C*	*B*	*D*	*A*
	4	*A*	*D*	*C*	*B*

Here each method of testing is used once a day, and once by each of the technicians. This design enables us to use pairwise comparisons of the testing procedures.

A Latin square design has its limitations. It is used to compare u treatment means when there are at most two extraneous sources of variability present. The influence of these sources is eliminated by special arrangement into u rows and u columns. The u treatments are randomly assigned to the rows and columns in such a way that each treatment appears in every row and every column of the Latin square design.

Problem Solving Example:

Q Four different bowlers, 1, 2, 3, and 4, use four different lanes, *A*, *B*, *C*, and *D*, on four different days, Sunday, Monday, Tuesday, and Wednesday. A Latin square is to be designed as follows:

	1	2	3	4
Sunday				
Monday				
Tuesday				
Wednesday				

Provide two different solutions for filling A, B, C, and D, where each letter must appear four times.

 The criteria for a Latin square are that each letter must appear only once in each column and each row.

Solution 1

	1	2	3	4
Sunday	A	D	C	B
Monday	B	A	D	C
Tuesday	C	B	A	D
Wednesday	D	C	B	A

Solution 2

	1	2	3	4
Sunday	A	D	C	B
Monday	C	B	A	D
Tuesday	B	A	D	C
Wednesday	D	C	B	A

CHAPTER 14

Count Data:
The Chi-Square Test

14.1 Count Data

In many situations, we gather the sample data measured on a quantitative scale. For example, to find the average yearly income of an American family, we choose a sample of families and measure their incomes. The result of each such measurement is the number of dollars earned. Data obtained from such experiments are called **count data**.

On the other hand, in some situations, the measurements are not quantitative. For example, we may want to classify students into three categories: good, average, and bad. We will be using the concept of a multinomial experiment.

14.1.1 Definition of Multinomial Experiment

1. The experiment consists of n identical trials.

2. There are k possible outcomes. Each trial result is one of k outcomes.

3. By p_i, $i = 1, ..., k$ we denote the probability that the outcome of a single trial will be i. This probability remains constant from trial to trial.

4. The outcome of one trial does not have any influence upon another trial.

5. The total number of trials resulting in outcome i is n_i. We have

$$\sum_{i=1}^{k} n_i = n.$$

Observe that since there are k outcomes

$$\sum_{i=1}^{k} p_i = 1.$$

The multinomial distribution is the probability distribution for the number of observations resulting in each of the k outcomes. It is given by the formula below.

$$P(n_1, n_2, \ldots, n_k) = \frac{n!}{n_1! \, n_2! \ldots n_k!} \, p_1^{n_1} p_2^{n_2} \ldots p_k^{n_k}$$

Problem Solving Example:

Q A package in the mail can either be lost, delivered, or damaged while being delivered. If the probability of loss is .2, the probability of damage is .1, and the probability of delivery is .7 and 10 packages are sent to Galveston, Texas, what is the probability that six arrive safely, two are lost, and two are damaged?

A If each package being sent can be considered an independent trial with three outcomes, the event of six safe arrivals, two losses, and two smashed packages can be assumed to have a multinomial probability. Thus,

$$Pr(6, 2, \text{ and } 2) = \frac{10!}{(6!)(2!)(2!)} \, (.7)^6 \, (.2)^2 \, (.1)^2 = .059.$$

The probability of six safe arrivals, two losses, and two damaged packages is .059.

14.2 Definition of χ^2

Let

$$E_1, ..., E_k$$

denote the set of possible events. The observed frequencies of events are

$$f_1, f_2, ..., f_k$$

respectively. Having the theoretical probabilities of events, we can compute the theoretical or expected frequencies of events

$$e_1, ..., e_k$$

respectively.

It is sometimes crucial to find out if the theoretical frequencies differ significantly from the observed frequencies.

For that purpose, we define the statistic χ^2

$$\chi^2 = \frac{(f_1 - e_1)^2}{e_1} + \frac{(f_2 - e_2)^2}{e_2} + ... + \frac{(f_k - e_k)^2}{e_k} \tag{1}$$

χ^2 measures the discrepancy between observed and expected frequencies.

If the total frequency is n,

$$\sum_{i=1}^{k} f_i = \sum_{i=1}^{k} e_i = n.$$

Equation (1) can be written as

$$\chi^2 = \sum_{i=1}^{k} {}_i \frac{f_i^2}{e_i} - n. \tag{2}$$

Indeed,

$$\chi^2 = \sum \frac{(f_i - e_i)^2}{e_i} = \sum \left(\frac{f_i^2 - 2f_ie_i + e_i^2}{e_i} \right)$$

$$= \sum \frac{f_i^2}{e_i} - 2\sum f_i + \sum e_i = \sum \frac{f_i^2}{e_i} - n$$

When $\chi^2 = 0$, then observed and expected frequencies agree exactly.

The sampling distribution of χ^2 is approximated closely by the χ^2 – distribution

$$y = y_0 \, x^{v-2} e^{-\frac{1}{2}\chi^2} \tag{3}$$

where v is the number of degrees of freedom.

Two situations occur:

1. The theoretical frequencies can be computed without the use of the population parameters estimated from sample statistics. Since

$$\sum_{i=1}^{k} e_i = n$$

we have to know only $k - 1$ of the expected frequencies.

$$v = k - 1$$

2. The theoretical frequencies can be computed only by m population parameters estimated from sample statistics.

$$v = k - 1 - m$$

Using the null hypothesis h_0, we compute the expected frequencies and then the value of χ^2. If the value of χ^2 is greater than some critical value, we conclude that observed frequencies differ significantly from expected frequencies and reject h_0.

Otherwise, we accept h_0.

This procedure is called the **chi-square test of hypothesis**.

EXAMPLE:

A coin was tossed 500 times. The result was 283 heads and 217 tails. Using the level of significance of 0.01, check the hypothesis that the coin was fair.

Observed frequencies of heads and tails are respectively

$$f_1 = 283$$

$$f_2 = 217$$

and expected frequencies are

$$e_1 = 250$$

$$e_2 = 250$$

We have

$$\chi^2 = \frac{(f_1 - e_1)^2}{e_1} + \frac{(f_2 - e_2)^2}{e_2}$$
$$= 4.356 + 4.356$$
$$= 8.712 \ .$$

There are two categories: heads and tails. Hence,

$$k = 2$$

and the number of degrees of freedom is

$$v = k - 1 = 1.$$

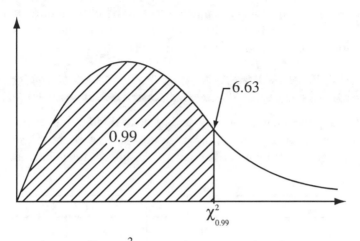

For $v = 1$ the value of $\chi^2_{0.99}$ is 6.63.

$$\chi^2_{0.99} = 6.63$$

The value of χ^2 is 8.712. Since

$$8.712 > 6.63$$

we reject the hypothesis that the coin is fair at a 0.01 level of significance.

Observe that the chi-square distribution is a nonsymmetrical distribution.

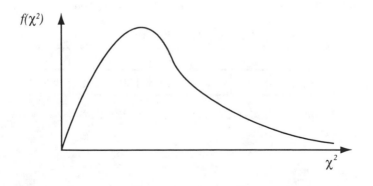

There are many chi-square distributions. We choose the appropriate one depending on the number of degrees of freedom.

The chi-square test is often used to determine the goodness of fit, that is, to determine how well theoretical distributions (such as binomial, normal, etc.) fit empirical distributions.

Problem Solving Examples:

 Suppose that a sales region has been divided into five territories, each of which was judged to have an equal sales potential. The actual sales volume for several sampled days is indicated below.

Territory	A	B	C	D	E
Actual Sales	110	130	70	90	100

What are the expected sales for each area? How many degrees of freedom does the chi-square statistic have? Compute χ^2 and use the table of χ^2 values to test the significance of the difference between the observed and expected levels of sales.

 If each sales area is judged to have equal sales potential, we would have expected each area to show an equal number of sales.

There were $110 + 130 + 70 + 90 + 100 = 500$ sales altogether and five sales regions. Thus, we would have expected $\frac{500}{5} = 100$ sales per region. The expected frequencies would be:

Territory	A	B	C	D	E
Expected Sales	100	100	100	100	100

There is one row and five columns; hence, there are $5 - 1 = 4$ degrees of freedom.

$$\chi^2 = \sum_{i=1}^{5} \frac{(O_i - E_i)^2}{E_i}$$

$$= \frac{(110-100)^2}{100} + \frac{(130-100)^2}{100} + \frac{(70-100)^2}{100}$$

$$+ \frac{(90-100)^2}{100} + \frac{(100-100)^2}{100}$$

$$= \frac{1}{100}[10^2 + 30^2 + 30^2 + 10^2 + 0^2]$$

$$= \frac{100 + 900 + 900 + 100}{100} = 20.0$$

The computed value of χ^2 with 4 degrees of freedom will be greater than 13.28 due to random factors only one time in 100 if the sales territories do in fact have equal sales potential.

Our calculated χ^2 statistic is 20 >13.28; thus, we see it is quite unlikely that the sales territories have equal sales potentials.

 A factory has four machines that produce molded parts. A sample of 500 parts is collected from each machine and the number of defective parts in each sample is determined:

Machine	1	2	3	4
Defects/500	10	25	0	5

Is there a difference between the machines? Outline an appropriate statistical test and, if possible, determine if these data are significant.

We use a chi-square test to determine if a significant relationship exists between the machine and the number of defective parts produced. The relevant hypotheses are:

H_O: There is no difference between machines.

H_A: There is a difference between machines.

The level of significance for this test will be $\alpha = .01$. We wish to determine the critical region of the test. The tower bound on the critical region will be the 99th percentile of the chi-square distribution with $4 - 1 = 3$ degrees of freedom. There are four classes; hence, the $4 - 1 = 3$ degrees of freedom.

Under the null hypothesis, H_O, the machines are equally likely to produce a defective item. If this is the case, then we would expect the total number of defectives, 40, to be equally divided between the four machines. That is, we would expect:

Machine	1	2	3	4
Defects/500	10	10	10	10

The chi-square statistic is

$$\chi^2 = \sum_{i=1}^{4} \frac{(\text{Observed} - \text{Expected})^2}{\text{Expected}}$$

$$= \frac{(10-10)^2}{10} + \frac{(25-10)^2}{10} + \frac{(0-10)^2}{10} + \frac{(5-10)^2}{10}$$

$$= 0 + 22.5 + 10 + 2.5$$

$$= 35$$

From the table of the chi-square distribution, we see that

$$Pr(\chi_3^2 > 11.34) = .01.$$

That is, the probability that a chi-square random variable with 3 degrees of freedom is greater than 11.34 is .01. Our statistic is greater than 11.34; thus, we reject H_O and accept the hypothesis that there is a significant relation between particular machines and the number of defects produced.

 A die was tossed 120 times and the results are listed below.

Upturned face	1	2	3	4	5	6
Frequency	18	23	16	21	18	24

(a) Compute the χ^2 statistic for this 1 by 6 contingency table under the hypothesis that the die was fair.

(b) Test the null hypothesis that the die is fair at the .05 level of significance.

 (a) The χ^2 statistic provides a means of comparing observed frequencies with expected frequencies.

This statistic is defined to be

$$\chi^2 = \sum_{i=1}^{4} \frac{(O_i - E_i)^2}{E_i}$$

where O_i is the observed frequency in cell i, E_i is the expected frequency found from some underlying probability model which must be assumed, and n is the number of cells.

The chi-square statistic (χ^2) is approximately distributed with a chi-square distribution. This distribution is found tabulated in many statistics texts. The χ^2 statistic has a parameter called degrees of freedom associated with it.

In this problem, we assume that the die is fair. That is, any face will land upturned with probability $\frac{1}{6}$. If this model is valid, we would expect equal numbers of each face to appear, or $\frac{120}{6} = 20$ occurrences of each face. The expected frequencies are

Upturned face	1	2	3	4	5	6
E_i, expected frequency if die is fair	20	20	20	20	20	20

Thus, we can compute the χ^2 statistic from the observed and expected frequencies.

The degrees of freedom of a 1 by n contingency table is $n - 1$. In our problem $n = 6$, thus $n - 1 = 6 - 1 = 5$ degrees of freedom.

The chi-square statistic will give some idea of the difference between the hypothesized model and the observed frequencies. The χ^2 statistic, with its distribution, will allow us to test the significance of the hypothesized probability model.

$$
\chi^2 = \sum_{i=1}^{6} \frac{(O_i - E_i)^2}{E_i}
$$

$$
= \frac{(18-20)^2}{20} + \frac{(23-20)^2}{20} + \frac{(16-20)^2}{20}
$$

$$
+ \frac{(21-20)^2}{20} + \frac{(18-20)^2}{20} + \frac{(24-20)^2}{20}
$$

$$
= \frac{1}{20}[(-2)^2 + 3^2 + (-4)^2 + 1^2 + (-2)^2 + 4^2]
$$

$$
= \frac{1}{20}[4+9+16+1+4+16] = \frac{50}{20}
$$

$$
= 2.5 \text{ with 5 degrees of freedom}
$$

(b) The critical chi-square value is 11.071, and since $2.5 < 11.071$, the null hypothesis is accepted.

14.3 Contingency Tables

Let

$$
E_1, E_2, ..., E_k
$$

be the set of possible events. The observed frequencies of events are

$$
f_1, ..., f_k
$$

correspondingly. The theoretical or expected frequencies are

$$
e_1, ..., e_k.
$$

Event	E_1	E_2	. . .	E_k
Observed frequency	f_1	f_2	. . .	f_k
Expected frequency	e_1	e_2	. . .	e_k

In the table, observed frequencies occupy one row. Such a table is called a **one-way classification table**. The number of columns is k, so the table is a

$$1 \times k \text{ table.}$$

There are tables where the observed frequencies occupy h rows. Such tables are called

$$h \times k \text{ tables.}$$

In general, these tables are referred to as **contingency tables**. To each observed frequency in an $h \times k$ table, there corresponds a theoretical frequency, usually computed according to the rules of probability. The contingency table consists of the cells.

Each cell contains a frequency called the cell frequency. The total frequency in each row or each column is called the marginal frequency.

The statistic

$$\chi^2 = \frac{\sum (f_i - e_i)^2}{e_i} \tag{4}$$

is applied to check agreement between observed and expected frequencies. The sum is taken over all cells in the contingency table. For $h \times k$ contingency tables this sum contains $h \times k$ terms. The sum of all observed frequencies is equal to the sum of all expected frequencies

$$\sum f_i = \sum e_i = n. \tag{5}$$

The statistic χ^2 defined by (4) has a sampling distribution approximated very closely by

$$y = y_0 \chi^{v-2} e^{-\frac{1}{2}\chi^2}. \tag{6}$$

Approximation gets better for large expected frequencies. The number of degrees of freedom v for

$$h > 1, \ k > 1$$

should be calculated as follows:

1. When the expected frequencies are computed without using population parameters estimated from sample statistics,

$$v = (h-1)(k-1) \tag{7}$$

2. When m estimated parameters are necessary to compute the expected frequencies,

$$v = (h-1)(k-1) - m \tag{8}$$

Contingency tables can be extended to higher dimensions. For example, we can design a table

$$h \times k \times l$$

where three classifications are present.

EXAMPLE:

Consider the 2×2 contingency table.

	I	II	Total
A	a_1	a_2	n_a
B	b_1	b_2	n_b
Total	n_1	n_2	n

Results Observed

Under a null hypothesis, we find the expected frequencies.

	I	II	Total
A	$\dfrac{n_1 n_a}{n}$	$\dfrac{n_2 n_a}{n}$	n_a
B	$\dfrac{n_1 n_b}{n}$	$\dfrac{n_2 n_b}{n}$	n_b
Total	n_1	n_2	n

Results Expected

Having observed and expected frequencies, we can compute χ^2

$$\chi^2 = \frac{\left(a_1 - \dfrac{n_1 n_a}{n}\right)^2}{\dfrac{n_1 n_a}{n}} + \frac{\left(a_2 - \dfrac{n_2 n_a}{n}\right)^2}{\dfrac{n_2 n_a}{n}}$$

$$+ \frac{\left(b_1 - \dfrac{n_1 n_b}{n}\right)^2}{\dfrac{n_1 n_b}{n}} + \frac{\left(b_2 - \dfrac{n_2 n_b}{n}\right)^2}{\dfrac{n_2 n_b}{n}}$$

(9)

But

$$a_1 - \frac{n_1 n_a}{n} = a_1 - \frac{(a_1 + b_1)(a_1 + a_2)}{a_1 + a_2 + b_1 + b_2} = \frac{a_1 b_2 - a_2 b_1}{n}.$$

Similarly, we find that

$$a_2 - \frac{n_2 n_a}{n} = \frac{a_1 b_2 - a_2 b_1}{n}$$

etc.

$$\chi^2 = \frac{n}{n_1 n_a}\left(\frac{a_1 b_2 - a_2 b_1}{n}\right)^2 + \frac{n}{n_2 n_a}\left(\frac{a_1 b_2 - a_2 b_1}{n}\right)^2$$

$$+ \frac{n}{n_1 n_b}\left(\frac{a_1 b_2 - a_2 b_1}{n}\right)^2 + \frac{n}{n_2 n_b}\left(\frac{a_1 b_2 - a_2 b_1}{n}\right)^2 \quad (10)$$

$$= \frac{n(a_1 b_2 - a_2 b_1)^2}{n_a n_b n_1 n_2}$$

EXAMPLE:

A new serum is tested for its effectiveness. Samples A and B, each consisting of 100 sick patients are tested. Group A used the serum and Group B did not use the serum.

	Recovered	Did Not Recover	Total
A (serum)	80	20	100
B (no serum)	60	40	100
Total	140	60	200

Frequencies Observed

The results are shown in the table. The null hypothesis h_0 states that the serum has no effect.

We calculate the expected frequencies under h_0 null hypothesis.

	Recovered	Did Not Recover	Total
A (serum)	70	30	100
B (no serum)	70	30	100
Total	140	60	200

Frequencies Expected under h_0

We have

$$\chi^2 = \frac{(80-70)^2}{70} + \frac{(60-70)^2}{70} + \frac{(20-30)^2}{30} + \frac{(40-30)^2}{30}$$
$$= 9.524.$$

Using the formula

$$\nu = (h-1)(k-1)$$

we find the number of degrees of freedom

$$\nu = 1.$$

Our results are $\chi^2 = 9.524$ for $\nu = 1$. For one degree of freedom

$$\chi^2_{0.995} = 7.88$$

The results are significant even at 0.005 level

$$7.88 < 9.524$$

We reject h_0 and conclude that the serum is effective.

14.3.1 Coefficient of Contingency

The degree of dependence or relationship of the classifications in a contingency table is measured by the coefficient of contingency defined as

$$\lambda = \sqrt{\frac{\chi^2}{\chi^2 + n}} \qquad (11)$$

The degree of dependence increases with the value of λ. Note that

$$0 \leq \lambda < 1.$$

If the number of rows and columns in the contingency table is equal to h, the maximum value of λ is

$$\lambda_{max} = \sqrt{\frac{h-1}{h}}.$$

EXAMPLE:

For the last example, the value of χ^2 was

$$\chi^2 = 9.524$$

and

$$n = 200.$$

The value of the coefficient of contingency is

$$\lambda = \sqrt{\frac{9.524}{9.524 + 200}} = 0.213.$$

14.3.2 Additive Property of χ^2

Repeated experiments are performed. For each one, the values of χ^2 and v are calculated.

$$\chi_1^2, \chi_2^2, \chi_3^2, \cdots$$
$$v_1, v_2, v_3, \cdots$$

The results of all these experiments are equivalent to one experiment with χ^2 given by

$$\chi^2 = \chi_1^2 + \chi_2^2 + \chi_3^2 + \cdots$$

and the number of degrees of freedom

$$v = v_1 + v_2 + v_3 + \cdots$$

EXAMPLE:

An experiment is performed four times in order to test hypothesis h_0. The values of χ^2 were 1.98, 3.14, 2.37, 1.90.

In each case, $v = 1$. At the level 0.05 for one degree of freedom, $v = 1$

$$\chi_{0.95}^2 = 3.84.$$

Thus, we cannot reject h_0 on the basis of any one experiment. Combining the results of the four experiments, we find

$$\chi^2 = 1.98 + 3.14 + 2.37 + 1.90 = 9.39$$

$$v = 1 + 1 + 1 + 1 = 4$$

and for $v = 4$

$$\chi^2_{0.95} = 9.49.$$

So, we can reject h_0 at the 0.05 level of significance.

Problem Solving Examples:

 Often frequency data are tabulated according to two criteria, with a view toward testing whether the criteria are associated. Consider the following analysis of the 158 machine breakdowns during a given quarter.

Number of Breakdowns

	Machine				
	A	*B*	*C*	*D*	Total per Shift
Shift 1	10	6	13	13	42
Shift 2	10	12	19	21	62
Shift 3	13	10	13	18	54
Total per Machine	33	28	45	52	158

We are interested in whether the same percentage of breakdown occurs on each machine during each shift or whether there is some difference due perhaps to untrained operators or other factors peculiar to a given shift.

 If the number of breakdowns is independent of the shifts and machines, then the probability of a breakdown occurring in the first shift and in the first machine can be estimated by multiplying the proportion of first shift breakdowns by the proportion of machine *A* breakdowns.

If the attributes of shift and particular machine are independent, then

Pr(breakdown on machine *A* during shift 1)

= *Pr*(breakdown on machine *A*) ×

Pr(breakdown during shift 1)

where Pr(breakdown on machine A) is estimated by

$$\frac{\text{number of breakdowns on } A}{\text{total number of breakdowns}} = \frac{33}{158}$$

and

Pr(breakdowns during shift 1)

$$= \frac{\text{number of breakdowns in shift 1}}{\text{total number of breakdowns}}$$

$$= \frac{42}{158}.$$

Of the 158 breakdowns, given independence of machine and shift, we would expect

$$(158)\left(\frac{42}{158}\right)\left(\frac{33}{158}\right) = \frac{42 \times 33}{158}$$

breakdowns on machine A and during the first shift.

Similarly, for the third shift and second machine, we would expect breakdowns.

$$\left(\frac{54}{158}\right)\left(\frac{28}{158}\right) \times 158 = \frac{54 \times 28}{158}$$

The expected breakdowns for different shifts and machines are

$$E_{11} = \frac{42 \times 33}{158} = 8.77,$$

$$E_{12} = \frac{28 \times 42}{158} = 7.44,$$

$$E_{13} = \frac{44 \times 42}{158} = 11.96,$$

$$E_{14} = \frac{52 \times 42}{158} = 13.82.$$

Similarly, the other expected breakdowns given independence are

	A	B	C	D
Shift 1	8.77	7.44	11.96	13.82
Shift 2	12.95	10.99	17.66	20.41
Shift 3	11.28	9.57	15.38	17.77

We will assume that the χ^2 test is applicable. There are $r = 3$ rows and $c = 4$ columns; thus, the chi-square statistic will have $(3-1)(4-1)$ = 6 degrees of freedom. The level of significance of this hypothesis will be $\alpha = .05$. This is the probability of rejecting independence of attributes (the null hypothesis) given that the attributes are in fact independent.

If this statistic is greater than 12.6, then we will reject the hypothesis that the machine and shifts are independent attributes in determining incidence of breakdown.

The chi-square statistic is

$$\chi_6^2 = \sum_{i=1}^{3} \sum_{j=1}^{4} \frac{(O_{ij} - E_{ij})^2}{E_{ij}} = 2.13.$$

We see that $2.13 < 12.6$. Thus, we accept that the attributes of machine and shift are independent in determining incidence of breakdown.

 Is there a significant relationship between the hair color of a husband and wife? A researcher observes 500 pairs and collects the following data.

WIFE	HUSBAND				Row Total
	Red	Blonde	Black	Brown	
Red	10	10	10	20	50
Blonde	10	40	50	50	150
Black	13	25	60	52	150
Brown	17	25	30	78	150
Column Total	50	100	150	200	500

The researcher wishes to test the theory that people tend to select mates with the same hair color.

Test this theory at the level of significance $\alpha = .01$.

 We wish to test the following hypotheses:

H_O: Husband's and wife's hair color are independent.

H_A: People tend to select mates with the same hair color. (Husband's and wife's hair color are dependent.)

We will now determine a critical region for our χ^2 statistic. The table has four rows and four columns, thus the statistic will have $(4-1)(4-1) = 9$ degrees of freedom. The critical region will be determined by a constant c such that if χ_9^2 is a random variable with 9 degrees of freedom,

$$Pr(\chi_9^2 > c) = .01.$$

From the table of the chi-square distribution, we see that $c = 21.67$. If our chi-square statistic is greater than 21.67, we will reject H_O and accept H_A. If the chi-square statistic is less than 21.67, we will accept H_O.

To compute the chi-square statistic, we first need to compute the expected frequencies based on a sample of 500 couples if the null hypothesis is true.

If the husband's and wife's hair colors are independent, the expected frequencies in the cells of the table will be determined by the marginal frequencies. We have to see that

$$E_{ij} = \frac{R_i - C_j}{n}$$

where R_i is the ith row total, C_j is the jth column total, and $N = 500$ is the total number of observations.

For example, the expected number of observations in cell $(4, 1)$, the expected number of couples where the husband has red hair and the wife has brown hair, is

$$\frac{50 \times 150}{500} = 15.$$

The expected frequencies for each cell are calculated in a similar way and are given below.

	H U S B A N D			
WIFE	Red	Blonde	Black	Brown
Red	5	10	15	20
Blonde	15	30	45	60
Black	15	30	45	60
Brown	15	30	45	60

The chi-square statistic is

$$\chi_9^2 = \sum_{i=1}^{4} \sum_{j=1}^{4} \frac{(O_{ij} - E_{ij})^2}{E_{ij}} = 32.57.$$

Since the chi-square statistic is greater than the critical value c, we reject the null hypothesis and accept the hypothesis that people tend to select mates with the same hair color.

14.4 Yates' Correction Factor: Some Helpful Formulas

Often, we apply the equations for continuous distributions to the sets of discrete data.

For the chi-square distribution, we use Yates' correction factor defined by

$$\chi_{corrected}^2 = \frac{\left(|f_1 - e_1| - 0.5\right)^2}{e_1} + ... + \frac{\left(|f_k - e_k| - 0.5\right)^2}{e_k}$$

The correction is made only when the number of degrees of freedom is one, $\nu = 1$.

Finally, we show how to compute χ^2 which involves only the observed frequencies.

2 x 2 Tables

	I	II	Total
A	a_1	a_2	n_a
B	b_1	b_2	n_b
Total	n_1	n_2	n

$$\chi^2 = \frac{n(a_1 b_2 - a_2 b_1)^2}{n_1 n_2 n_a n_b}$$

$$= \frac{n(a_1 b_2 - a_2 b_1)^2}{(a_1 + b_1)(a_2 + b_2)(a_1 + a_2)(b_1 + b_2)}$$

With Yates' correction

$$\chi^2_{\text{corrected}} = \frac{n\left(\left| a_1 b_2 - a_2 b_1 \right| - \frac{n}{2}\right)^2}{(a_1 + b_1)(a_2 + b_2)(a_1 + a_2)(b_1 + b_2)}$$

$$= \frac{n\left(\left| a_1 b_2 - a_2 b_1 \right| - \frac{n}{2}\right)^2}{n_1 n_2 n_a n_b}$$

2 x 3 Tables

	I	II	III	Total
A	a_1	a_2	a_3	n_a
B	b_1	b_2	b_3	n_b
Total	n_1	n_2	n_3	n

We can use the general formula

$$\chi^2 = \sum \frac{f_i^2}{e_i} - n$$

to find

$$\chi^2 = \frac{n}{n_a}\left(\frac{a_1^2}{n_1} + \frac{a_2^2}{n_2} + \frac{a_3^2}{n_3}\right) + \frac{n}{n_b}\left(\frac{b_1^2}{n_1} + \frac{b_2^2}{n_2} + \frac{b_3^2}{n_3}\right) - n.$$

The Yates' Correction Factor should be used when the table of values has any of the following conditions:

1. two rows and two columns.

2. one row and two columns.

3. all entries for expected frequencies are under 5, regardless of the number of rows and columns. (This is a rarity.)

Problem Solving Examples:

Q At the annual shareholders' meeting of the Syntho-Diamond Company, the managing director said he was pleased to report that they had captured a further six percent of the gem market. "How do you know?" asked a shareholder. The managing director replied "Last year we conducted a survey of 1,000 people who owned jewelry. They were selected quite haphazardly, so that we believe them to be a random sample. Of this group, 400, that is 40%, owned one or more of our artificial diamonds. Last week we conducted another such survey, and found 460 out of 1,000 owning our gemstones; this represents a rise to 46%." This didn't impress the shareholder, who retaliated, "It looks to me as though the difference might easily be due to chance." The managing director was able to answer this doubt within a few minutes. Can you?

 A First formulate a contingency table representing the results of the survey:

	Owned Artificial Diamonds	Owned Real Diamonds	Row Totals
First Random Sample	400	600	1,000
Second Random Sample	460	540	1,000
Column Totals	860	1,140	2,000

This table represents the results of two surveys.

Because this contingency table has only two rows and two columns, the chi-square statistic must be corrected.

This correction factor is designed to correct for the approximation of a discrete distribution by a continuous distribution. To make this correction, we compute the χ^2 statistic in the following way.

Let O_{ij} = observed frequency in cell ij and E_{ij} = expected frequency in cell ij.

Let $Y_{ij} = |O_{ij} - E_{ij}| - .5$.

Then the Yates' χ^2, as this corrected chi-square statistic is called, is

$$\chi^2 = \sum_{j=1}^{2} \sum_{i=1}^{2} \frac{Y_{ij}^2}{E_{ij}}.$$

The degrees of freedom for this statistic will be $(2-1)(2-1) = 1$.

If the observed frequencies in this table are due to chance, then we would expect half of the 860 respondents to be in each survey.

Thus, the expected frequencies are

430	570
430	570

The Yates' chi-square statistic is computed in the table below.

O_{ij}	E_{ij}	$Y_{ij} = |O_{ij} - E_{ij}| - .5$	$\dfrac{Y_{ij}^2}{E_{ij}}$
400	430	$29.5 = |430 - 400| - .5$	2.02
600	570	$29.5 = |600 - 570| - .5$	1.53
460	430	$29.5 = |460 - 430| - .5$	2.02
540	570	$29.5 = |570 - 540| - .5$	1.53

Thus, $\chi_1^2 = 7.10$.

The chance of observing this statistic due to chance alone is found from the table of chi-square values. The χ^2-statistic will be greater than 6.63 only one percent of the time. Since $7.10 \geq 6.63$, there appears to be a significant increase in the company's share of the gem market.

 Suppose a new cancer medicine, called Treatment Q134, is developed and tested on a random sample of 170 patients. Of these patients, 150 are cured and 20 die.

An almost identical group of 170 patients are tested with the older treatment Q133 and of these patients 130 are cured and 40 die.

Test the significance of these results.

 The data is represented in the contingency table below.

Number of Patients

Treatment	Cured	Died	Row Totals
Q134 (new)	150	20	170
Q133 (old)	130	40	170
Column Totals	280	60	340

If one treatment is not significantly better than the other, then we would expect the same number of patients to die and to be cured in each sample of 170. Thus, we might expect:

	Cured	Died
Q134	140	30
Q133	140	30

We now conduct a significance test to see if the observed frequencies are significantly different from those expected. We will compute the χ^2 statistic from this data using Yates' correction factor to $O_{ij} - E_{ij}$ because the table has two rows and two columns and the total number of observations, $n = 340$, is large.

If the resulting χ^2 statistic is greater than some constant c, where $Pr(\chi^2 \geq c) \leq .05$, then we will accept the significance of the relationship between the treatment used and the number of patients cured. The level of significance will be .05.

The constant c is chosen from the tabulated values of the χ^2 distribution. Because our statistic is computed from a table with two rows and two columns, the statistic will have $(r-1)(c-1) = (2-1)(2-1) = 1$ degree of freedom.

From the tables of the chi-square distribution, we see that the value of c such that $Pr(\chi^2 > c) = .05$, for a χ^2 random variable with 1 degree of freedom is $c = 3.84$.

To compute the chi-square test statistic, we use the following table.

$E_{ij} =$ Expected Frequency of Cell i, j	$O_{ij} =$ Observed Frequency of Cell i, j	$Y_{ij} = \|O_{ij} - E_{ij}\| - .5$	$\dfrac{Y_{ij}^2}{E_{ij}}$
140	150	$10 - .5 = 9.5$	.65
140	130	$10 - .5 = 9.5$	.65
30	40	$10 - .5 = 9.5$	3.0
30	20	$10 - .5 = 9.5$	3.0

$$\chi^2 = \sum_{j=1}^{2} \sum_{i=1}^{2} \frac{Y_{ij}^2}{E_{ij}} = .65 + .65 + 3.0 + 3.0 = 7.3$$

Our χ^2 statistic is greater than $c = 3.84$; so there is a significant difference between the two treatments.

On the basis of this test, we should begin to use the new treatment on cancer patients of this type.

CHAPTER 15

Time Series

15.1 Time Series

A time series is a set of measurements taken at specified times. Usually, the measurements are taken at equal time intervals. For example, the Gross National Product (GNP) is measured every year, or every quarter, or every month.

We measure the value of y (income, temperature, etc.) at times

$$t_1, t_2, \ldots$$

Here, y is a function of t:

$$y = y(t).$$

15.2 Relatives and Indices

A simple way of comparing a series of observations is to use a ratio.

EXAMPLE:

Monthly sales volumes for a six-month period are known.

Month	1	2	3	4	5	6
Sales	$20,000	$16,000	$24,000	$31,000	$29,000	$37,000

Monthly Sales Volumes

We want to compare sales in each month with the sales in a specified month, say the first month. In this case, month 1 is the base period. We divide each month's sales by sales in the first month.

Month	1	2	3	4	5	6
Ratio	1.00	0.80	1.20	1.55	1.45	1.85
%	100	80	120	155	145	185

Ratios and Relatives

Usually, the ratios are multiplied by 100 to give percentages. The results are called **relatives**.

Linked relatives show each period as a percentage of the preceding period. Thus, month 1 is used as the base period for month 2; month 2 is used as the base period for month 3; etc.

Month	1	2	3	4	5	6
Sales	$20,000	$16,000	$24,000	$31,000	$29,000	$37,000
Linked Relatives		80	150	129	94	128

Linked Relatives

Linked relatives show data for each period as a percentage of the preceding period.

15.2.1 Indices

Often, to describe how the price of food or total production outcome changes over a period of time, we construct an index (which is a single number) that reflects the changes over a period of time.

A good example is a food price.

EXAMPLE:

Over five years, the prices of the following items were recorded.

Prices	Year				
	1	2	3	4	5
Milk ($/gallon)	1.68	2.05	2.01	2.23	2.42
Potatoes ($/lb)	0.72	0.64	0.81	0.80	0.86
Beef ($/lb)	1.65	2.12	1.97	2.32	2.75

To construct an index, we can use a simple aggregate. That is, we add up all the prices and compute relatives based on their sum. Using year 1 as the base period, we find:

Year	1	2	3	4	5
	1.68	2.05	2.01	2.23	2.42
	0.72	0.64	0.81	0.80	0.86
	+1.65	+2.12	+1.97	+2.32	+2.75
Total	4.05	4.81	4.79	5.35	6.03
Index	100	119	118	132	149

While computing the simple aggregate, we assume that all elements are equally important. It is not always the case.

To remedy the situation, we use a weighted aggregate index, where weights are assigned arbitrarily or are based on some system of preferences.

Problem Solving Examples:

 Plot the following two time series on the same graph and compare.

Coal production Tons (millions)	Year	Wage earners in industry Number (thousands)
210.8	1	706.2
211.5	2	701.8
207.4	3	697.4
198.8	4	692.7
183.9	5	602.1
176.8	6	531.0

A There is no natural relationship between the units of measurement and thus no obvious basis of comparison. We resolve this difficulty with the use of index numbers. The index numbers used here are the percentage changes in each year of the series as compared with 1952. For example, the index number for coal production in 1954 is

$$\frac{211.5}{210.8} \times 100 = 100.3$$

	Year	Coal Production Index	Wage Earners Index
Similarly,	1	$\frac{210.8}{210.8} \times 100 = 100.0$	$\frac{706.2}{706.2} \times 100 = 100.0$
	2	$\frac{211.5}{210.8} \times 100 = 100.3$	$\frac{701.8}{706.2} \times 100 = 99.4$
	3	$\frac{207.4}{210.8} \times 100 = 98.4$	$\frac{697.4}{706.2} \times 100 = 98.8$
	4	$\frac{198.8}{210.8} \times 100 = 94.3$	$\frac{692.7}{706.2} \times 100 = 98.1$
	5	$\frac{183.9}{210.8} \times 100 = 87.2$	$\frac{602.1}{706.2} \times 100 = 85.3$
	6	$\frac{176.8}{210.8} \times 100 = 83.9$	$\frac{531.0}{706.2} \times 100 = 75.2$

We now have a basis of comparison. For each category we place the years on the *x*-axis and index numbers on the *y*-axis. We plot the points and connect them with straight lines.

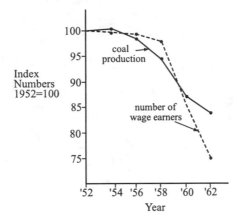

 Consider the following table which shows prices over a four year period.

	Year			
Prices	1	2	3	4
Milk per gallon	$1.80	$1.95	$2.00	$2.21
Potatoes per lb.	$0.68	$0.72	$0.75	$0.90
Beef per lb.	$2.10	$2.33	$2.80	$3.05

Using year 1 as the base year, construct an index number for each year.

 The sums for years 1, 2, 3, and 4 are $4.58, $5.00, $5.55, and $6.16, respectively. Now assign 100 to year 1.

$$\text{Index for year 2} = \frac{5.00}{4.58}(100) = 109$$

$$\text{Index for year 3} = \frac{5.55}{4.58}(100) = 121$$

$$\text{Index for year 4} = \frac{6.16}{4.58}(100) = 134$$

If 1990 is the base year, it automatically gets assigned an index of 100. The index for 1991 = (sum of 1991 prices ÷ sum of 1990 prices) × 100 and likewise for the other years' indices.

15.3 Analysis of Time Series

Time Series $y = y(t)$ is often represented by a graph.

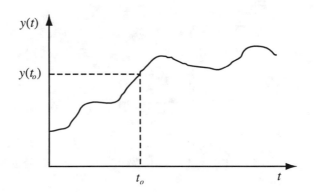

Analysis of time series shows certain characteristic measurements or variations, which all time series exhibit to varying degrees.

We classify characteristic movements of time series into four categories called **components of a time series**.

1. Long-Term Movement

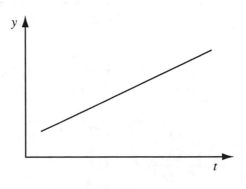

Long-Term Trend

Long-term movements refer to the general direction in which the graph of a time series is moving over a long period of time. For example, a long-term increase in sales volume as a result of population growth is a **trend**.

2. Cyclical movements, or cyclical variations, refer to the long-term oscillations about a trend curve. These cycles are not necessarily periodic.

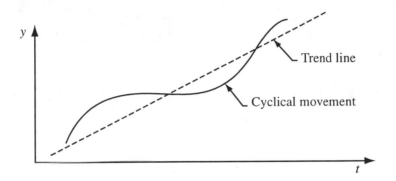

Cyclical Movement

3. Seasonal movements, or seasonal variations, refer to patterns, identical or almost identical, which a time series follows during corresponding months of successive years. Month-by-month changes in sales related to holidays and changes in the weather are examples of seasonal variation.

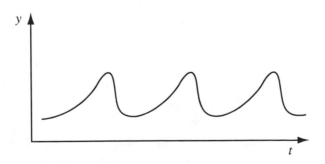

Seasonal Variation

4. Random movements refer to the motions of time series due to chance events such as war, strikes, etc. Usually, it is assumed that such variations last only a short time.

In analyzing time series, we assume that the variable y is a product of the variables T (trend), C (cyclical), S (seasonal), and R (random) movements.

$$y = T \times C \times S \times R$$

The analysis of time series consists of decomposition of a time series into its basic components, T, C, S, and R, which are then analyzed separately.

15.4 Moving Averages

A set of numbers is given

$$y_1, y_2, y_3, \ldots$$

We define a moving average of order n to be the sequence of arithmetic means

$$\frac{y_1 + \ldots + y_n}{n}, \frac{y_2 + \ldots + y_{n+1}}{n}, \frac{y_3 + \ldots + y_{n+2}}{n}, \ldots$$

EXAMPLE:

In the sequence of natural numbers

$$1, 2, 3, 4, 5, 6, 7, \ldots$$

a moving average of order 3 is given by the sequence

$$\frac{1+2+3}{3} = 2, \frac{2+3+4}{3} = 3, \frac{3+4+5}{3} = 4,$$

$$\frac{4+5+6}{3} = 5, \ldots$$

which is the sequence of natural numbers starting with the number 2.

If data are supplied annually or monthly, we talk about an n year moving average or n month moving average.

In general, moving averages tend to reduce the amount of variation in a set of data.

In the case of time series, it is called smoothing of time series. We eliminate unwanted fluctuations.

15.4.1 Trends

There are several methods of estimation of trend.

1. The method of least squares is used to find an appropriate trend line or trend curve.

2. Fitting a trend line or trend curve by looking at the graph is called the freehand method.

3. We can eliminate the cyclical, seasonal, and irregular patterns by using moving averages of the proper orders. What is left is the trend movement. This method is called the moving average method.

4. We can separate the data into two parts, possibly equal, and average the data in each part. The two points obtained define a trend line. This method is called the method of semi-averages.

Problem Solving Example:

Q Consider the following table giving daily ticket sales figures for a movie theater.

	Week 1	Week 2
Sunday	900	800
Monday	400	500
Tuesday	500	300
Wednesday	600	300
Thursday	300	400
Friday	700	600
Saturday	1,100	900

Compute a seven-day moving average.

A We proceed in the following manner. First, find the average daily attendance from Sunday to Saturday of the first week. This figure is placed next to Wednesday. Then find the average from Monday of the first week to Sunday of the second week. Proceeding in this way we obtain the seven-day moving average for the given data. Thus,

	Week 1 Daily Sales	7 day M.A.	Week 2 Daily Sales	7 day M.A.
Sunday	900		800	571
Monday	400		500	586
Tuesday	500		300	571
Wednesday	600	643	300	543
Thursday	300	629	400	
Friday	700	643	600	
Saturday	1,100	614	900	

Examining the original data, we see that sales fluctuate. On weekends, the number of patrons is larger than on weekdays. Note now that these fluctuations are reduced by the moving average method. The moving average typically reduces seasonal (or in this case weekly) variability. Thus, for example, assume we have a series of monthly sales figures extending over a number of years. If we wish to identify the trend (i.e., reduce the seasonal variations), one way of doing so is to use the method of moving averages. This method involves fewer computations than the use of seasonal index adjustments.

Quiz: Experimental Design – Time Series

1. Six students were given a standardized exam in math and English, with results shown in the chart that follows.

Student	Amy	Bob	Cathy	Don	Edna	Frank
Math Score	80	70	65	90	85	60
English Score	75	60	65	80	75	70

In testing the null hypothesis of no correlation between math scores and English scores on this exam, which of the following statements is correct?

(A) Reject the null hypothesis at the .05, .02, and .01 levels of significance.

(B) Accept the null hypothesis at the .05 level, but reject it at the .02 and .01 levels.

(C) Not enough information is given to determine any correlation between the math scores and English scores.

(D) Accept the null hypothesis at the .05 and .02 levels, but reject it at the .01 level.

(E) Accept the null hypothesis at the .05, .02, and .01 levels of significance.

2. A farmer found that the coefficient of correlation between the amount of rainfall during the first three weeks of planting their crop and the yield per acre was 0.76. This means that

(A) 76% of the variations in crop yield can be explained by variations in rainfall during the first three weeks.

(B) 76% of the variations in rainfall during the first three weeks can be explained by variations in crop yield.

(C) 58% of the variations in crop yield can be explained by variations in rainfall during the first three weeks.

(D) 58% of the variations in rainfall during the first three weeks can be explained by variations in crop yield.

(E) 76% of the time, crop yield is dependent on the amount of rainfall during the first three weeks.

3. A researcher performs an experiment, changing the value of the independent variable x and recording values of the dependent variable y. She calculates the value of the correlation coefficient to be $r = 0.7$, and draws the associated regression line. What percentage of the variation in y cannot be attributed to changes in x and the associated relationship indicated on the regression line?

(A) 70%

(B) 49%

(C) 30%

(D) 51%

(E) There is not enough information provided.

4. Which of the following are good reasons for blocking data in an experiment involving multiple measures?

I.   To reduce within-subject variability
II.   To make statistical analysis more manageable
III. To aid interpretation of the results of subsequent analyses
IV.   To increase the reliability of the data

(A) I and II   (D) I and III

(B) II and III   (E) II and IV

(C) III and IV

5. A marketing firm is interested in studying the potential market for a new cereal. The cities of Boston, Chicago, Detroit, and Miami will be studied as potential markets. Four different kinds of packaging P_1, P_2, P_3, and P_4 will be used; four kinds of advertis-

ing A, B, C, and D will be used. For the following Latin square, how should the letters A, B, C, and D be placed in the third column from top to bottom?

	P_1	P_2	P_3	P_4
Boston	A	B		D
Chicago	B	C		A
Detroit	C	D		B
Miami	D	A		C

(A) ADBC   (D) CDBA

(B) DBCA   (E) CDAB

(C) ADCB

6.   A teacher wishes to study the effects of administering mathematics, science, and history tests in different orders, with a five-minute break between tests. He assembles three groups of students in different rooms —Group 1, Group 2, and Group 3. Which of the following arrangements would be a faulty design for the study?

I.

	First	**Second**	**Third**
Mathematics	Group 3	Group 2	Group 1
Science	Group 1	Group 3	Group 2
History	Group 2	Group 1	Group 3

II.

	First	**Second**	**Third**
Mathematics	Group 1	Group 2	Group 3
Science	Group 3	Group 1	Group 2
History	Group 2	Group 3	Group 1

III.

	First	**Second**	**Third**
Mathematics	Group 1	Group 2	Group 2
Science	Group 3	Group 3	Group 1
History	Group 2	Group 1	Group 3

IV.

	First	**Second**	**Third**
Mathematics	Group 3	Group 2	Group 1
Science	Group 1	Group 3	Group 2
History	Group 2	Group 1	Group 3

V.

	First	**Second**	**Third**
Mathematics	Group 2	Group 1	Group 3
Science	Group 1	Group 3	Group 2
History	Group 3	Group 2	Group 1

(A) I   (D) IV

(B) II   (E) V

(C) III

7. Using the goodness of fit test, an investigator wants to show that people have clear preferences among ten different brands of laundry detergents. If he surveys 100 people, which of the following would prove his point at the 0.01 level of significance?

(A) $\chi^2 > 135.8$ (D) $\chi^2 > 21.67$

(B) $\chi^2 > 23.21$ (E) $\chi^2 > 140.2$

(C) $\chi^2 < 135.8$

8. A sociologist is studying the relationship between church attendance and donations to charity. He sets up a table as follows:

Attendance	Large Donations	Average Donations	Small Donations
Often (more than once a week)			
Average (once a week)			
Occasional (between average & never)			
Never			

What would prove at the 0.05 level of significance that the amount of a donation is related to church attendance?

(A) $\chi^2 > 12.59$ (D) $\chi^2 > 11.07$

(B) $\chi^2 > 14.06$ (E) $\chi^2 > 16.01$

(C) $\chi^2 > 21.02$

9. The table below shows the number of people from different age ranges enrolled in day and evening programs in a small community college. Assuming that there was no relationship between age and choice between day and evening classes, how many people over 30 years of age should be enrolled in evening programs?

	Evening Program	Day Program	Total
Under 25 years	400	850	1,250
25 to 30 years	475	750	1,225
31 years & over	560	300	860
Total	**1,435**	**1,900**	**3,335**

(A) 633   (D) 370

(B) 490   (E) 300

(C) 400

10. The following table shows the numbers of women and men under 35 years old, by age group, serving as volunteer firefighters in one city. The fire chief wishes to test the null hypothesis that there is no significant difference between the distributions of male and female volunteers in the various age categories. Which of the following figures most closely approximates the chi-square values associated with the difference between the observed and the expected frequencies in this table?

Distribution of Volunteer Firefighters by Gender

	18-24	25-29	30-34
Female	17	23	30
Male	22	40	36
Total	39	63	66

(A) 6.149   (D) 1.026

(B) 1.590   (E) 0.443

(C) 1.138

ANSWER KEY

1.	(E)	6.	(C)
2.	(C)	7.	(D)
3.	(D)	8.	(A)
4.	(A)	9.	(D)
5.	(E)	10.	(C)